AN INTRODUCTION TO BIRDS

AN INTRODUCTION TO BIRDS

By
H.S. Bhamrah
Kavita Juneja

Edited by
Balvinder K.

ANMOL PUBLICATIONS PVT. LTD.
NEW DELHI - 110 002 (INDIA)

ANMOL PUBLICATIONS PVT. LTD.
4374/4B, Ansari Road, Daryaganj
New Delhi - 110 002

An Introduction to Birds

First Edition, 1990
Reprint, 1991, 1992, 1994
Second Revised Edition, 2001

ISBN 81-261-0703-0

PRINTED IN INDIA

Published by J.L. Kumar for Anmol Publications Pvt. Ltd., New Delhi - 110 002 and Printed at Mehra Offset Press, Delhi.

Contents

Preface

We feel pleasure in bringing out the second revised edition of the present book on the instance of our learned teachers. Besides rectifying the errors of first edition, new material is added here and there in the light of recent development and researches in the field.

This text is designed to approach the morphology, anatomy, physiology and development of the *Birds* in a coherent way since it is most natural for a student in studying the architectural elements of an animal body in the quest to know how they function. This text intends, therefore, to enable the student, through his own natural appetency, to correlate form and function so that he possesses for subsequent work in either anatomy or physiology a firm foundation for a real edifice of knowledge.

The book features both a text and a laboratory guide. It deals with the structures and physiological phenomena of each system. Origin, adaptative radiation and other general topics are also dealt in details. It is hoped that this book will not only meet the requirement of Indian students but will also be useful as a guideline to the teachers in their teaching.

There can be no claim to originality except in the manner of treatment and much of the information has been obtained from the books and scientific journals available in different libraries.

We express our thanks to our friends and colleagues whose constant inspiration have initiated them to bring out this book.

The authors are also grateful to Dr. J. Bahadur, Professor and Head, School of Studies in Zoology, Jiwaji University Gwalior, Dr. A.K. Pandey, Professor & Head, Vikram University, Ujjain for their valuable suggestions.

All the constructive criticisms and suggestions for improvement from learned teachers and scholars are welcome.

H.S. Bhamrah
Kavita Juneja

1

Aves : Characters and Classification

CHARACTERS

Aves constitute a well-defined group of vertebrates. They possess a senes of strongly marked characters such as distinguish hardly any other class. The distinguished characters of birds are-

1. Body is more or less spindle shaped and divisible into small head, neck, short trunk and tail.
2. Bipedal; fore-limbs are modified as wings provided with feathers for flight. Hind-limbs are large, strong and adapted for perching, walking, swimming etc. and bear claws.
3. Skin is thin and devoid of glands except *uropygial* present at the base of tail. Epidermal exoskeleton is in the form of feathers, claws, beaks and a few scales on the hind limbs.
4. Endoskeleton is fully ossified, light but strong and without epiphysis. Long bones are pneumatic or hollow containing extension of air-sacs from lungs, thus bone-marrow is absent.
5. Skull *monocondylic* i.e. with single occipital condyle.
6. Jaws without teeth and modified into *beaks*.
7. Vertebral column is short. Vertebrae with *heterocoelous* centra. A *synsacrum* is present and is formed by the fusion of posterior thoracic, lumbar, sacral and anterior caudal vertebrae.
8. Sternum is keeled for the attachment of flight muscles. Clavicles along with interclavicles form *furcula* - a wish bone.
9. Ribs double headed with backwardly directed uncinate process.
10. Carpals and metacarpals fused to form *carpometacarpus* similarly distal rows of tarsals and metatarsals unite into *tarsometatarsus*. The proximal row of tarsals fuses with tibia to form *tibio-tarsus.*
11. Heart is completely 4-chambered. Right aortic (systemic) arch is present, however, left arch disappears in adult. Renal portal system vestigial , however, well developed hepatic portal system present.

12. Erythrocytes are oval, nucleated and biconvex.
13. *Homoiothermous* i.e. warm blooded animals.
14. Respiratory system is much complicated and provided with air-sacs, Lungs are spongy instead of hollow.
15. Larynx with out vocal cords. A sound box or *syrinx* lies at or near the junction of trachea and bronchi.
16. Kidneys are metanephric. Birds are urecotelic i.e., uric acid is the waste product. Urinary bladder is absent.
17. Brain is large but smooth. It has two most prominent optic lobes.
18. 12 pairs of cranial nerves.
19. Eyes are large provided with *pecten*.
20. Olfactory organs poor. Middle ear contains a single ossicle.
21. Sexes are separate Female with only left ovary and oviduct i.e. right ovary and oviduct are vestigial. Fertilization is internal and takes place in oviduct. It is preceded by copulation and courtship.
22. Eggs are large with large amount of yolk (*megalecithal egg*), covered by calcareous shell.
23. Eggs are developed by external incubation. Newely-hatched young may be fully formed (precocial or indifugos) or immature (altricial or nidiculous).

About 9000 living species of birds are known at present. According to *Wetmore* (1960) there are 34 orders, 27 orders of living birds of which two have recently become extinct and 7 orders of fossil birds.

CLASSIFICATION

Class-Aves is divided into two sub-classes.

Sub-class. I-Archaeornithes

1. Primitive birds of upper Jurassic period lived about 155 million years ago and became extinct.
2. Wings primitive with little power of flight.
3. Small cerebellum.
4. Skull with teeth on both the jaws.
5. Vertebrae amphicoelous.
6. Sternum without keel.
7. Carpals and metacarpals free.
8. Long tail with 18-20 free caudal vertebrate. Pygostyle or fused-tail vertebrae absent.

9. Ribs without uncinate process.
10. Abdominal ribs absent.

This sub class includes a single order.

Order-Archaeopterygiformes

Examples; *Archaeopteryx lithographica* and *Archaeornis*.

Sub-class. II-Neornithes

1. Including modern as well as few post Jurassic extinct birds.
2. Wings usually well-developed and adapted for flight with few exceptions.
3. Cerebellum large.
4. Teeth are absent except in few fossil birds.
5. Digits of the fore limbs are without claws.
6. Vertebrae heterococlous in living forms.
7. Tail is short with few free caudal vertebrae. Pygostyle present.
8. Sternum usually with a keel.
7. Ribs with uncinate processes
8. Abdominal ribs are absent.
9. Carpometacarpus present.

This sub-class is divided into 4 super orders.

Super-order-(A) Odontognathae

1. Extinct, upper Cretaceous birds.
2. Jaws with teeth.
3. Brain of avian type.
4. Pygostyle absent.

It is divided into two orders.

Order-1. Hesperornithiformes

1. Large, flightless marine birds.
2. Vertebrae amphicoelous.
3. Pleurodont teeth.
3. Pectoral girdle reduced..
4. Sternum without a keel.

Examples; *Hesperornis, Baptornis, Enaliornis* etc.

Order-2. Ichthyornithiformes

1. Marine birds.
2. Neck vertebrae amphicoelous.

3. Presence of teeth is doubtful.
4. Pectoral girdle well developed.
5. Sternum with a well developed keel.

Examples: *Ichthyornis, Apatornis.*

Super order-(B) Ratitae or Palaeognathae

1. Modern, large-sized, flightless, running birds.
2. Wings are reduced, vestigial or absent.
3. Feathers devoid of interlocking mechanism.
4. Rectrices or tail feathers are absent or irregulary arranged.
5. Sternal keel vestigial, absent or flat.
6. Free caudal vertebrae are present. Pygostyle is small or absent.
7. Uncinate process of ribs vestigial or absent.
6. Clavicles are small or absent.
7. Ilium and ischium not united posteriorly except in *Emu* and *Rhea*.
8. Syrinx or sound box is absent.
9. Male has a large and erectile penis, female has clitoris.
10. Youngs are precocious.

The flightless birds are not represented in India, Super-order-Ratitae is divided into 7-orders.

Order-1. Struthioniformes

1. Head, neck and legs are sparsely feathered.
2. Beak is short and broad
3. Legs strongly developed each with two toes with stunted nails.
4. Sternum without keel.
5. Pubes form a ventral symphysis.

Example: Ostriches *(Struthio camelus),*

Order-2. Rheiformes

1. Flightless terrestial and swift running birds.
2. Head, neck and thigh are feathered.
3. Legs bear 3 toes with heavy and typical claws.
4. Sternum without keel.
5. Ischia form a ventral symphysis.

Example: Darwin's Rhea, *(Pteroncemia pennata), and* American ostrich *(Rhea americana)*

Order-3. Casuariformes

1. Large terrestrial flightless birds.

2. Head bears a comb-like structure.
3. Neck and body are densly feathered.
4. Feathers with aftershaft
5. Fore limbs are greatly reduced..
6. Three toes with typical claws on each foot are present.
7. Sternum without keel.

   Examples: Cassowaries (*Casurius*) of Austraiia and New Guinea and Emus *(Dromaius novaehollandiae*) of New Zealand.

Order-4. Apterygiformes

1. Flightless terrestrial birds.
2. Feathers simple, hair-like or bristle like. Feathers without aftershaft.
3. Wings are vestigial.
4. The beak is long and slender with nostrils at the top of the maxillae.
5. Long legs with 4 toes
6. The eyes are small.
7. Sternum without keel.

   Example : Kiwi *(Apteryx*) of New Zealand.

Order-5. Dinornithiformes

1. Giant flightless terrestrial birds, became extinct nearly 700 years ago.
2. Feathers with large aftershaft.
3. Reduced fore-limbs. Hind-limbs massive with 3 or 4 toes.
4. Beak short and massive.
5. Reduced sternum without keel.
6. Pectoral girdle completely absent.

   Example : Moas *(Dinornis maximus)* of New Zealand.

Order-6. Aepyornithiformes

1. Flightless terrestrial extinct elephant-bird recently became extinct.
2. Long powerful legs with 4 toes.
3. Wings are vestigial.
4. Sternum is without keel.

   Examples: *Aepyornistitan, Mulleornis.*

Order-7. Tinamiformes

1. Small terrestrial birds, cursorial with poor power of flight..
2. Wings well developed but essentially great runners.
3. Tail feathers and pygostyle are reduced.

4. Sternum keeled.

   Examples: Tinamou *(Tinamus), Eudromia.*

Super order-(C) Impennae

1. Modern aquatic, birds .
2. Flightless birds
3. Adapted to aquatic life
4. Feet are webbed
5. Feathers small, scale-like.
6. Wings paddle-like.

   It includes only one order. Order Sphenisciformes.

   Example: Penguins *(Aptenodytes)* of Southern Hemisphere.

Super order-(D) Carinatae or Neognathae

1. Mostly modern, usually small-sized flying birds, distributed all over the world. Wings well developed.
2. Feathers with interlocking mechanism.
3. Skull is neognathous, that is, vomer is short allowing palatines to meet.
4. Sternum keeled.
5. Quadrate is double headed.
6. Teeth absent, beak present.
7. Rectrices (tail-feathers) present and arranged regularly.
8. Oil gland is present.
9. Pygostlyle present. Scapula and coracoid meet at a right angle or acute angle.
10. Clavicles are well developed.
11. Flight muscles are well developed.
12. Young are altricious. Male has no copulatory organ.

    Super order- carinatae includes twenty two (22) orders.

    These are as given below :

Group (A) Arboreal Birds

Order-1 Passeriformes

1. This is the largest of all the bird orders including half the known species.
2. Three toes are infront and one behind.
3. Feet are adapted for perching, while beaks are adapted for cutting.

Examples: Common house sparrow *(Passer domesticus)*, common house crows *(Corvus splendens)*, common myna *(Acridotheres tristis)*, weaver bird (*Ploceus philippines),* bulbul *(Molpestes);* starling (*Sturnus*) etc.

Order-2 Piciformes

1. Beak is strong chisel-shaped.
2. Tongue is long, flat horny and protrusible.
3. Tail feathers are stiff with pointed tips.
4. Insectivorous. Bore holes in trees.
5. Two toes infront and are behind, not reversible.

Examples: Wood pecker *(Picus),* yellow wood pecker *(Dendrocopos)*, golden beak wood pecker *(Dinopium)*

Order-3 Columbiformes

1. Skin thick and soft.
2. Bill or beak is short and slender
3. Crop is large and secretes 'pigeon milk' to feed young ones.
4. Graminivours and frugivorous.

Examples: Blue rock pigeon *(Columba livia)*, doves and extinct dodo *(Raphus).*

Order-4. Psittaciformes

1. Feathers are green, blue, yellow or red.
2. Two toes infront and two behind.
3. Strongly bent beak, fleshy tongue. Upper jaw is movably articulated with the cranium.
4. Frugivorours birds.

Examples: Large Indian Parakeet *(Psittacula eupatria)*, green parrot *(P. krameri)* etc.

Group (B) Terrestrial Birds

Order-5. Galliformes

1. Terrestrial game birds with short and powerful flight.
2. Feet are adapted for perching.
3. Feather with aftershaft.
4. Bill is short.
5. Graminivorous.

Examples: Peacock (*Pavo cristatus*), red jungle fowl (*Gallus*), quail (*Cotrunix)* and pheasants *(Phasianus)*

Order-6. Cuculiformes

1. Two toes infront and two toes behind.
2. Feet not adapted for grasping.
3. Tail long and beak moderate.
4. Many cuckoos are parasitic, the female lays eggs in the nest of crows.

   Examples: Cuckoo *(Cuculs canorus)*, koel *(Eudynamis scolopaceous)* etc.

Group (C) Swimming and diving birds

Order-7.Anseriformes

1. Tail is usually short consisting of many feathers.
2. Aquatic, anterior toes are webbed.
3. Beak is also broad and depressed, and is covered with soft cornified epidermis..
4. Tongue is fleshy.
5. Body is also covered with soft cornified epidermis.

   Examples: Swan (*Cygnus*), goose *(Anser)* etc.

Order-8. Coraciiformes

1. Long keeled angular beak.
2. Beak is strong.
3. Anterior toes are united and are adapted for diving.
4. Legs are short.
5. Fish catchers.

   Examples : Pied kingfishers (*Ceryle*), white breasted kingfisher (*Halcyon*) horn-bills (*Tochus*), hoopoe (*Upupa*) etc.

Order-9. Gaviiformes

1. Marine birds with webbed feet.
2. Legs are short and at the end of the body.
3. Powerful in flight.
4. Tail comprises 18-20 short stiff-feathers.

   Example : Loons *(Gavia)*

Order-10. Colymbiformes or Podicipediformes

1. Fresh water divers.
2. Legs are placed far back on the body.
3. Tail comprises a tuft of down feathers.

4. Tarsus is compressed.
   Examples : *Podicipes* and *Podilymbus*

Order-11. Procellariformes

1. Birds with great power of flight.
2. Skull with large nasal glands.
3. Toes are webbed.
4. Feathers compact and oily in texture.
5. Wings are long and narrow.
6. Stictly marine, nest usually on islands.
   Examples : Albatrosses (*Diomedea*), Petrels (*Procellaria*) etc.

Order-12. Pelcaniformes

1. Aquatic birds with webbed toes.
2. Piscivorous i.e., fish-eater.
3. Nostrils are vestigial or absent.
4. Presence of gular pouch on the throat.
5. Beaks are long with wide gaps for catching and swallowing the prey.
   Examples: Pelicans *(Pelecanus),* Indian darter (*Anhinga*)

Group (D) Shore birds and wading birds

Order-13. Charadriiformes

1. Shore dwelling aquatic birds.
2. Fore and hind limbs are slender and enormously elongated.
3. Toes are webbed at base
4. Mud probing beaks.
5. Feathers are dense and firm.
   Examples : Sand piper *(Tringa glariola*), gull *(Larus*), curlew *(Numenius*)

Order-14. Ciconiiformes

1. Large birds with long neck and long legs.
2. Feathers are decorative.
3. Feet are webbed type.
4. Beaks abruptly decurved.
   Examples : Cattle egret *(Bubulcus ibis*), heron *(Ardea herodias*), stork *(Ciconia)* flamingo *(Phonicopterus).*

Order-15. Gruiformes

1. Marshy Birds with heavy beaks.

2. Legs and toes are long and partly webbed.
3. Feathers with aftershaft.

   Examples : Sarus crane (*Antegone antegone*), bastard (*Choriotis*), common coot (*Fulica atra*).

Group (E) Birds of prey

Order-16. Falconiformes

1. Carnivorous birds.
2. Mandibles with sharp edged.
3. Curved beak which is hooked at the tip.
4. Diurnal and strong fliers.
5. Toes are powerfully clawed and feet are modified for perching.

   Examples: Eagle *(Aquila)*, hawk *(Falco)*, vultures *(Sarcogyps)*, kite *(Milvus migrans)*.

Order-17. Strigiformes

1. Nocturnal in habit.
2. Head large and rounded.
3. Strongly hooked beak is bent downwardly from base.
4. Ear opening large, often with flap-like cover.
5. Eyes are large directed forward and surrounded by circle of stiff feathers.

   Examples: Great horned owl *(Bobo bubo)*, brown fish owl *(Ketupa zeylonesis)*

Group (F) Aerial Birds

Order-18. Micropodiformes or Apodiformes

1. Usually small birds.
2. Legs are short and wings are large.
3. Beak is long and oval shaped.
4. Insectivorous.

   Exampes: Indian swifts *(Micropodus)*, palm swifts *(Cypsiurus)* and humming birds.

Order-19. Caprimulgiforemes

1. Nocturnal and insectivorous.
2. Beak is short, flat and triangular often beset with stiff bristtes.
3. Mouth is wide and margined with long brsitle like sensory feathers.
4. Feathers are soft.

5. **Legs and feet small and weak, not adapted for grasping.**
   Examples : Night hawks (*Chordeilies*), goat suckers (*Caprimulgus*).

ARCHAEOPTERYX

Phylum	-	Chordata
Group	-	Vertebrata
Sub-phylum	-	Gnathostomata
Class	-	Aves
Sub-class	-	Archaeornithes
Genus	-	*Archaeopteryx*

Fig. 1.1. Fossil *Archaeopteryx*.

Its two specimens have been obtained from the Jurassic deposits of Bavaria in S. Germany. The first fossil was discovered by *Andreas Wagner* (1861). These fossils are connecting link between reptiles and birds because they show the mixture of both.

Reptilian Characters

1. Presence of teeth on jaws.
2. Bones are non-porous and non-hollow.
3. Lizard-like tailed body with about 20 caudal vertebrae which do not fuse into pygostyle.
4. The shape and mode of articulation of vertebrae like that of reptiles and centrum is amphicoelous.
5. Ribs are single headed without uncinate process.
6. Abdominal ribs are present similar to that of *Sphenodon* and Crocodile.
7. Sternum may be weak or absent.
8. The comparative length of the fore-and hind-limbs is of reptilian type. Skeleton of hand presents reptilian pattern. Phalanges being 2, 3 and 4 respectively in first, second and third digits which terminating into a claw.
9. Metacarpals are separate, carpometacarpus is absent.
10. Elongated ilium and backwardly directed pubis in the pelvic girdle.
11. Sclerotic plates are present around the eyes.

Avian Characters

1. Presence of feathers on tail, and fore-limbs.
2. Fore-limbs are modified into wings.
3. Skull bones are compactely fused.
4. Presence of 'V' shaped furcula bone or "merry thought" bone.
5. Cranium is large with large orbtis and single occipital condyle (monocondylic skull).
6. Jaws drawn out into a bill or beak.
7. The pubic bone in pelvic girdle is found in backwardly directed condition.
8. Scapula slender and curved.
9. Foot consists of tarsometatarsus and four clawed digits.
10. The opposable condition of the first toe of feet.

STRUTIO CAMELUS

Phylum	-	Chordata
Group	-	Vertebrata
Sub-phylum	-	Gnathostomata
Class	-	Aves
Sub-class	-	Neornithes
Super-order	-	Palaeognathae or Ratitae
Order	-	Struthioniformes
Genus	-	*Struthio*
Species	-	*camelus*

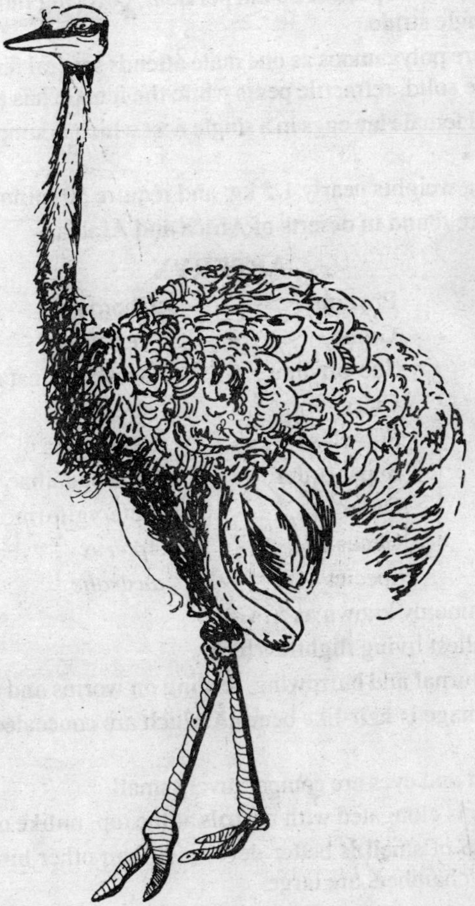

Fig. 1.2. *Struthio camelus* (Ostrich).

1. It is commonly called as *Camel bird* and in Hindi, *Shutura Murg*, and it is a giant flightless bird.
2. It is one of the largest living bird measuring 8 feet in hight and about 300 lbs in weight.
3. Head, neck and legs are sparsely feathered. Feathers are plume-like and without aftershafts.
4. Male and female birds differ in the colour of their plumage.
5. Neck is considerably long.
6. The feet are hooves like and there are only two toes i.e., 3rd and 4th. The inner toe is quite large and strong.
7. The food consists of herbage including fruits and seeds.
8. It can run at a speed of 80 km per hour, covering more than 8 meters in a single stride.
9. They are polygamous as one male attends several females. Male has a single solid, retractile penis while the female has a clitoris.
10. Several female lay eggs in a single nest which is simple hollow in the sand.
11. The egg weights nearly 1.5 kg. and require 50 minutes to boil it.
12. They are found in deserts of Africa and Arabia.

APTERYX

Phylum	-	Chordata
Group	-	Vertebrata
Sub-phylum	-	Gnathostomata
Class	-	Aves
Sub-class	-	Neornithes
Super-order	-	Palaeognathae
Order	-	Apterygiformes
Genus	-	*Apteryx*
Species	-	*australis*

1. It is commonly known as *Kiwi*.
2. It is smallest living flightless bird.
3. It is nocturnal and burrowing, feeding on worms and insects.
4. The plumage is hair-like beneath which are concealed rudimentary wings.
5. The head and eyes are comparatively small.
6. The beak is elongated with nostrils at the top, unlike other birds.
7. The sense of smell is better developed than other birds, hence the olfactory chambers are large.

Fig. 1.3. *Apteryx.*

8. Skull is pneumatic but other bones are solid.
9. Sternum with keel.
10. Pelvic girdle is simple with pubis and ischium free for their entire length, the pubis has a distinct pectineal process below the acetabulum.
11. It is a fast runner and runs by jumping legs with three toes in front and the hallux behind.
12. The male has a penis and female has a clitoris. Female lays two eggs in a burrow which are large and incubated by males.
13. The kiwi are found only in New Zealand.

ANTENODYTES

Phylum	-	Chordata
Group	-	Vertebrata
Sub-phylum	-	Gnathostomata
Class	-	Aves
Sub-class	-	Neornithes
Super-order	-	Impennae
Genus	-	*Antenodytes*

1. They are commonly known as *Penguins*.
2. They are flightless and marine birds.
3. The fore-limbs are modified into flippers and are well developed.
4. The hind-limbs are much posteriorly placed on the trunk and are use as oars.
5. The toes of feet are webbed with a lather like skin as in ducks.

Fig. 1.4. *Antenodytes.*

6. The skin is fully covered with feathers which are spear-shaped. The shaft is flat.
7. They feed upon fishes, crustaceans, molluscans. The young ones are feed by the parents with half digested crustaceans from their gullet.
8. The female lays only one egg which is placed in the folds of feet web.
9. They are economically important as oil and guano is obtained from them.
10. They are found from Dassen Islands of South Africa to Antarctica.

BUBO

Phylum	-	Chordata
Group	-	Vertebrata
Sub-phylum	-	Gnathostomata
Class	-	Aves
Sub-class	-	Neornithes
Super-order	-	Neognathae
Order	-	Strigiformes
Genus	-	*Bubo*
Species	-	*bubo*

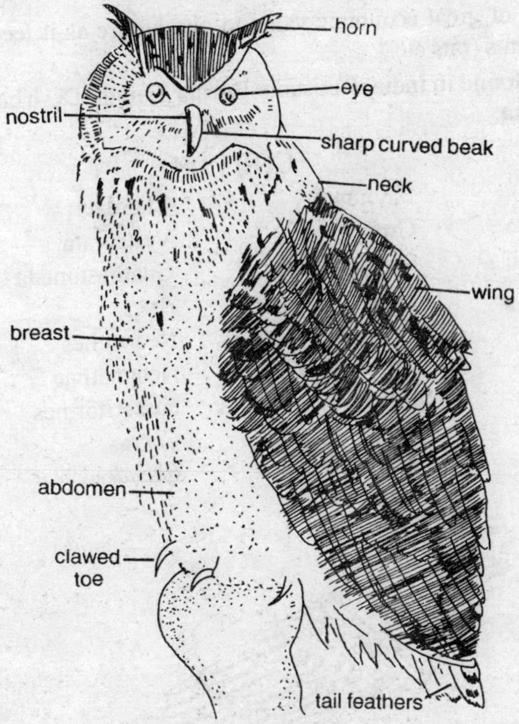

Fig. 1.5. *Bubo bubo.*

1. It is commonly known as the 'great horned owl' and in Hind as *ghughu* or *ulloo*.
2. It is large heavy and robust bird measuring up to 60 cm in length.
3. The colour is dark brown streaked and mottled with buff and black.
4. Head is quite large with large, rounded yellow and forwardly directed eyes.
5. Two prominant feather tufts above the head look like ears or horns.
6. Beak is short and parrot like.
7. Feet are adapted for grasping with large, sharp and curved, claws.
8. It is nocturnal bird as it hides during day time under bushes, rock projections etc.
9. It is carnivorous feeds on small birds, mammals, reptiles and occasionally on fishes and crabs.
10. The nest is on bare soil under a bush. Female lays 3-4-creamy eggs.

11. It is of great economic value to agriculture as it feeds on harmful rodents (rats etc.).
12. It is found in India, Pakistam, Burma, Bangla-Desh but absent in Sri Lanka.

CORVUS

Phylum	-	Chordata
Group	-	Vertebrata
Sub-phylum	-	Gnathostomata
Class	-	Aves
Sub-class	-	Neornithes
Super-order	-	Neognathae
Order	-	Passeriformes
Genus	-	*Corvus*
Species	-	*splendens*

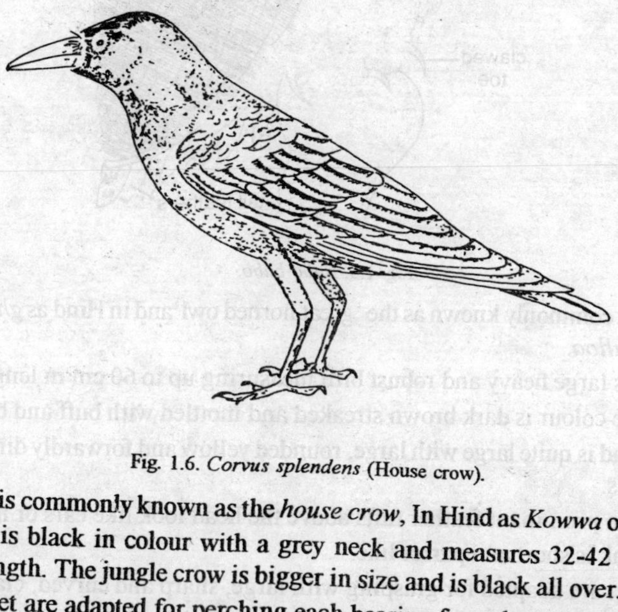

Fig. 1.6. *Corvus splendens* (House crow).

1. It is commonly known as the *house crow*, In Hind as *Kowwa* or *Kag*.
2. It is black in colour with a grey neck and measures 32-42 cm in length. The jungle crow is bigger in size and is black all over.
3. Feet are adapted for perching each bearing four clawed toes, three directed in front and one behind.
4. It is omnivorous and eats each and everything, but has a special liking for the food man eats that is why the crows spend most of their day time near human dwellings and spend night in the company of others crows far away.

5. They are intelligent, clever and cunning. They do not breed in company. Each pair makes its nest apart. The nest looks ugly and untidy from outside but it is neat, tidy and cosy from inside.
6. Female lays four eggs in May which are dull bluish green colour blotched with brown.
7. They are scavengers. They destroy locusts and other insects when they are swarming. But they also damage crops and fruit orchards.
8. They are distributed throughout India, Pakistan, Sri Lanka and Burma.

MILVUS

Phylum	-	Chordata
Group	-	Vertebrata
Sub-phylum	-	Gnathostomata
Class	-	Aves
Sub-class	-	Neornithes
Super-order	-	Neognathae
Order	-	Falconiformes
Genus	-	*Milvus*
Species	-	*migrans*

Fig. 1.7. *Milvus migrans* (Pariah kite).

1. It is commonly called *pariah kite*. In Hind as *Cheel*.
2. It is large bird about 60 cm in length.

3. Scavanger feeds on rodents, reptiles, garbage, insects, worms offal and other things which can be picked up.
4. The plumage is reddish brown streaked with dark brown. There is no sexual dimorphism.
5. The tail is forked. Beak is sharp, pointed and hooked.
6. Nesting season is from September to April. The nest is untidy platform of sticks and rubbish of every kind 3-4 pinkish white eggs are laid in the nest. Eggs are incubated by both the parents.
7. It is found in India, Burma, Sri Lanka and Pakistan.

EUDYNAMIS

Phylum	-	Chordata
Group	-	Vertebrata
Sub-phylum	-	Gnathostomata
Class	-	Aves
Sub-class	-	Neornithes
Super-order	-	Neognathae
Order	-	Cuculiformis
Genus	-	*Endynamis*
Species	-	*Scolopaceous*

Fig. 1.8. *Eudynamis scolopaceous.* Koel (male and female).

1. It is commonly called as *Koel*.
2. Body is slender with a longer tail measuring up to 17 inches in length.
3. It is a common bird of gardens and grooves, is perhaps better known by its voice than its appearance.
4. It feeds on figs of banyan and peepal, various barries and hairy catterpillers.

5. Sexual dimorphism is well marked, the male is characterized by having glistening metallic black all over the body with yellowish-green beak and small blood red eyes. The female is dull brown with white spots and lines.
6. The call 'Kuoo' is produced by males only. In winters it is silent.
7. It never builts its own nest. Between April to August the female bird goes in search of a nest of crow, where it lays its eggs. The eggs are pale grey with brown speckeles and look very much like those of crow.
8. It is distributed in India, Burina, Sri Lanka and Pakistan.

PSITTACULA KRAMERI

Phylum	-	Chordata
Group	-	Vertebrata
Sub-phylum	-	Gnathostomata
Class	-	Aves
Sub-class	-	Neornithes
Super-order	-	Neognathae
Order	-	Psittaciformes
Genus	-	*Psittacula*
Species	-	*krameri*

1. It is commonly called as *rose-ringed parakeet* and *Tota* in Hindi.
2. Body is slender with a long pointed tail and measures about 10 inches in length.
3. Body is covered by a brilliant grass green plumage.
4. Beak is short but stout, sharp edged deeply hooked and coral red in colour.
5. There is a sexual dimorphism, the female is green all over but the male has a rose pink collar and a black throat.
6. Feet are adapted for grasping, holding and climbing.
7. There are two front toes and two hind toes.
8. It is frugivorous as feeds on fruits.
9. They are gregarious with loud voice.
10. Birds live in holes of trees or buildings and the female lays 2-4 white blunt oval eggs. The eggs are incubated by both the parents.
11. It is a popular cage bird and can be taught to repeat few words and sentences.
12. It is found throughout India, Pakistan, Sri Lanka and Burma.

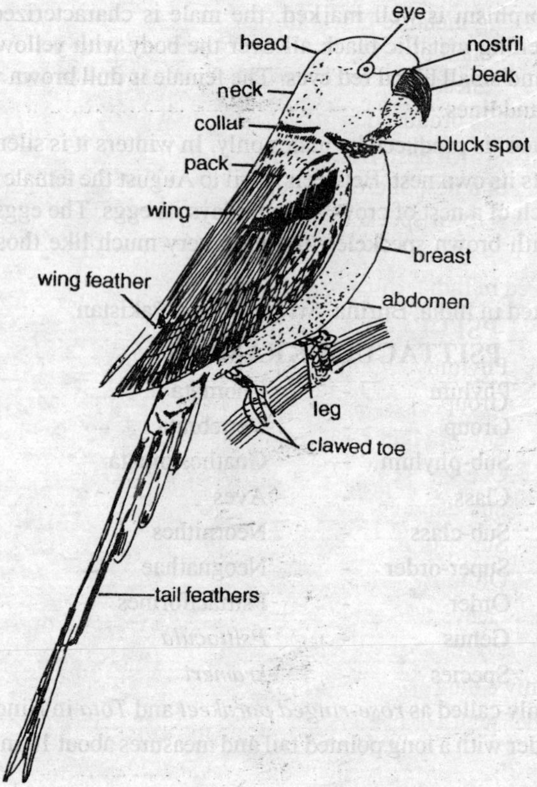

Fig. 1.9. *Psittacula.*

PASSER DOMESTICUS

Phylum	-	Chordata
Group	-	Vertebrata
Sub-phylum	-	Gnathostomata
Class	-	Aves
Sub-class	-	Neornithes
Super-order	-	Neognathae
Order	-	Passeriformes
Genus	-	*Passer*
Species	-	*domesticus*

1. It is commonly known as *house sparrow* and in Hindi it is called as *chidiya* or *gauriya*.
2. It is a small bird measuring 10-16 cm in length.

Fig. 1.10. *Passer domesticus.* Male house sparrow.

3. They are brown in colour with white and black mingled together on the back and sides, and whitish front.
4. Sexual dimorphism is seen. The male has a black patch on the throat and breast.
5. Sparrows are omnivorous, but chiefly granivorous feeds on grains. They destroy leafly vegetables and flower buds. They are useful too as they destroy several agricultural insect pests.
6. Nest is a collection of straw and rubbish stuffed into a hole in a wall, on roofs etc.
7. Breeding occurs throughout the year. Female lays 3 to 5 pale greenish white eggs in the nest. The incubation period is 14 days. Newly hatched young ones are cared by parents.
8. They are present all over the world (cosmopolitan) except Andaman and Nicobar.

ACRIODOTHERES TRISTIS

Phylum	-	Chordata
Group	-	Vertebrata
Sub-phylum	-	Gnathostomata
Class	-	Aves
Sub-class	-	Neornithes
Super-order	-	Neognathae
Order	-	Passeriformes
Genus	-	*Acriodotheres*
Species	-	*tristis*

Fig. 1.11. *Acriodotheres.*

1. It is commonly known as *Myna*.
2. It measures about 22 cm in length. It is seen every where in towns, villages, fields and garden.
3. The colour is perky dark brown getting paler on the under-side. Head, throat and chest are glossy black. Beak and legs are yellow.
4. There is a small patch of bare yellow skin behind the eyes.
5. There is no sexual dimorphism.
6. They are omnivorous feeding on fruits, insects etc.
7. The nest is made up of twigs, roots, papers, rubbish etc. in a hole of a tree or wall. Female lays 4 or 5 glossy blue eggs. Both the parents incubate them.
8. Hill myna (*Gracula religiosa*) is the best talker and mimic amongst the Indian birds.
9. It is found through-out India, Pakistan.

DINOPIUM BENGHALENSE

Phylum	-	Chordata
Group	-	Vertebrata
Sub-phylum	-	Gnathostomata
Class	-	Aves
Sub-class	-	Neornithes
Super-order	-	Neognathae
Order	-	Piciformes
Genus	-	*Dinopium*
Species	-	*benghalense*

1. It is commonly known as *woodpecker*. In Hindi it is called *Kathphorwa*.
2. It is small bird of about 20 cm in length.

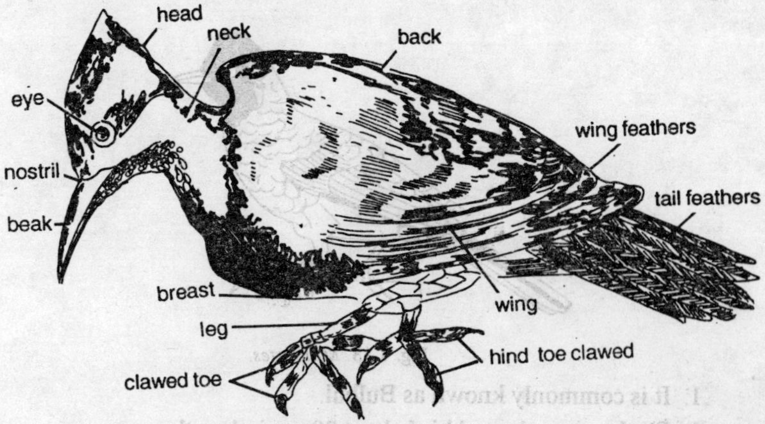

Fig. 1.12. *Dinopium.*

3. Upper plumage is distinctive golden yellow and black while lower is buffy white streaked with black.
4. The crown and occipital crest are more crimson in male than the female.
5. The bill or beak is chisel like, square at the tip and with an edge always kept sharp.
6. Tongue is long and protrusible possessing sharp and strong hooks or barbs at the tip.
7. Four toes are present in feet, 2 infront and 2 behind.
8. They live in pairs and feed on the beetles, ants, grubs and other insects thus they are insectivorous.
9. Nest is formed in a tree hole excavated by them. Female lays three white eggs between March and August which are incubated by both the sexes.
10. They are found in India, Pakistan and Sri Lanka.

MOLPASTES CAFER

Phylum	-	Chordata
Group	-	Vertebrata
Sub-phylum	-	Gnathostomata
Class	-	Aves
Sub-class	-	Neornithes
Super-order	-	Neognathae
Order	-	Passeriformes
Genus	-	*Molpastes*
Species	-	*cafer*

Fig. 1.13. *Molpastes.*

1. It is commonly known as Bulbul.
2. It is brown coloured bird about 20 cm in length.
3. The head is black, scale-like markings are present on beak and breast and a crimson patch at the root of tail.
4. Sexual dimorphism is not seen.
5. It feeds on insects, vegetables and flower nectar.
6. The nest is formed in a bush or tree. The nest is a cup of rootlets sometimes plastered with cobwebs on the outside.
7. Eggs are laid in the month of February to May. Their number is 2 or 3 and pinkish white in colour. Both the sexes incubate the eggs.
8. It occurs in India, Bangla Desh, Burma, Sri Lanka. In Pakistan it is rare or absent at all.

Revision Questions

1. Classify Aves giving important characters and example of each group.
2. Write short notes on : (a) *Archaeopteryx,* (b) *Bubo,* (c) Ostrich and (d) Myna.

2

Pigeon (Columa livia)

There are many domestic varieties of pigeons which differ from one another in size, proportions, colouration and details of feather arrangement. All these have been produced from blue rock pigeon, *Columba livia* by animal breaders through artificial selections. The following description refers to the *Columba livia.*

SYSTEMATIC POSITION

Phylum	-	Chordata
Group	-	Vertebrats
Sub-phylum	-	Gnathostomata
Class	-	Aves
Sub-class	-	Neornithes
Order	-	Columbiformes
Genus	-	*Columba*
Species	-	*livia*

Habits and Habitat

The blue rock pigeon lives in rocky and inaccessible places like caves, fissures and holes of rocks, forts, old ruins but never nests on trees. Although wild, they refers to live in those places of towns and cities which have large amount of grains like grain markets, temples, railway stations, churches, tombs, mosques etc.

The pigeons are inoffensive, harmless, gregarious birds. They commonly feed on grains, pulses, seeds of fruits and grasses. Sometimes they are found to eat insects, snails and slugs, probably mistaken for seeds. It dips its entire beak under water and drinks like a horse instead of sipping like other birds.

It never sings, chips or screeches. Its sound is a deep *gootur goon, gootur goon*. It is incapable of rapid locomotion on land because of short legs and rather weak feet. The mode of walking on two legs is called *bipedal gait*. The flight is swift.

The sexes are alike. Polyandry and polygamy are unknown among them. They lead a *monogamous* life i.e. one male lives and copulates with only one female throughout the life. They are also very devoted to their children.

Breeding season lasts all the year round in temperate countries at least. It displays a brief courtship before copulation. Fertilization is internal followed by copulation. Male and female incubate the eggs for fourteen to eighteen days. The immature, helpless and featherless young ones are fed by both the sexes on pigeon's, milk. Which is a whitish fluid secreted by their crops during the breeding season.

Enemies

In nature there are many enemies which attack them. The common predators are cats, foxes, hawks, weasels, owls and snakes. Men also prey on them for flesh and study.

Economic importance

They are useful as well as harmful. They provide easy food to various predators. They form nice pets in households and important animals for study in laboratories. They have been loved to be a reliable messangers during *First World War*. Besides they consume large quantities of grains.

Distribution

The blue rock pigeon is widely distributed in Asia, Europe and North Africa. They are plentiful in Palestine City of Israel. In India there are two sub-species of *Columba livia*. One is *Columba livia neglecta* which is large, paler in colour and is common up to 13,000 feet on the Himalayas. The other is *Colimba livia intermedia* which is smaller and darker and occurs throughout India.

EXTERNAL MORPHOLOGY

Shape, Size and Colouration

The body is streamlined or spindle shaped to offer minimum resistance to wind during flight. It measures about 32-33 cm in length. Except feet, eyes and beak which are usually pink, rest of the body is blue-grey colour with a beautiful glistening, blue green and purple sheen on the upper breast and around the neck. The wings posses two dark bars and a band across the end of tail.

Divisions of body

The body is bilaterally symmetrical and divided into four regions, the *head*, *neck*, *trunk* and *tail*.

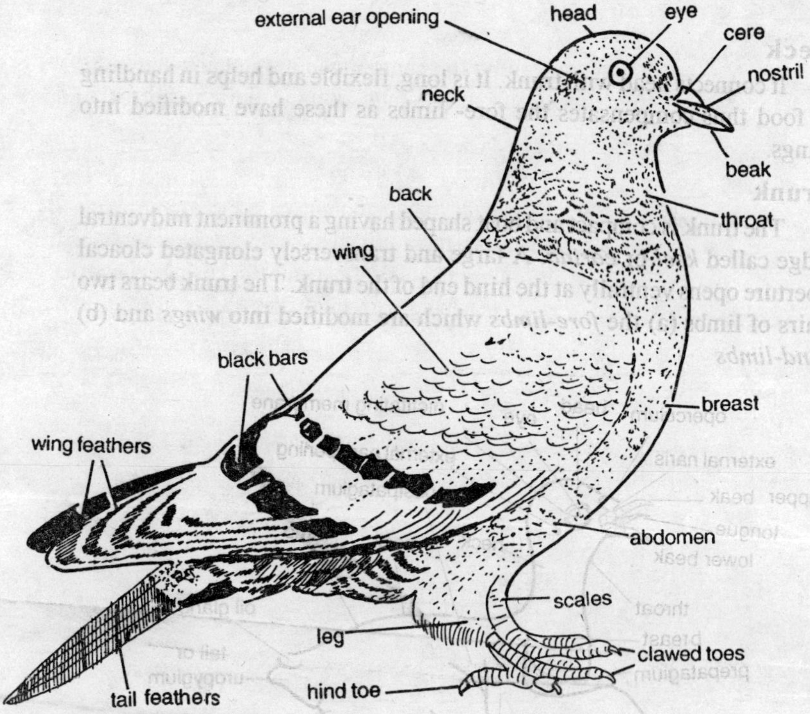

Fig. 2.1. *Columba livia*

Head

It is the anteriormost part of the body which is small, rounded and mobile. Infront it bears a short, straight and pointed *beak* or *bill* formed by the elongation of upper and lower jaws. It is adapted for picking up grains only. Both the jaws of beak are devoid of teeth and externally remain covered with a horney sheath called *rhamphotheca*. The mouth is terminal and has a wide gape. At the base of upper beak there are two slit-like apertures called *external nares* or *nostrils*. Each nostril is dosally overhung by a patch of naked, whitish, swollen area of soft skin called *cere* or *operculum*. The head bears two large *eyes* on lateral sides. The eyes are oval and each is guarded by an *upper* and a *lower eyelid* and a transparent third eyelid or *nictitating membrane*. The nictitating membrane can be pulled across the eyeball from the anterior corner of the eye. The eyelids are bare without eyelashes and eyebrows. Behind the eyes are small rounded apertures called *ears* or *auditory openings* which are covered with *auricular feathers*.

Neck

It connects head with trunk. It is long, flexible and helps in handling of food thus compensates the fore- limbs as these have modified into wings.

Trunk

The trunk is compact and boat shaped having a prominent midventral ridge called *keel* or *carina*. A large and transversely elongated cloacal aperture opens ventrally at the hind end of the trunk. The trunk bears two pairs of limbs (a) the *fore-limbs* which are modified into *wings* and (b) *hind-limbs*

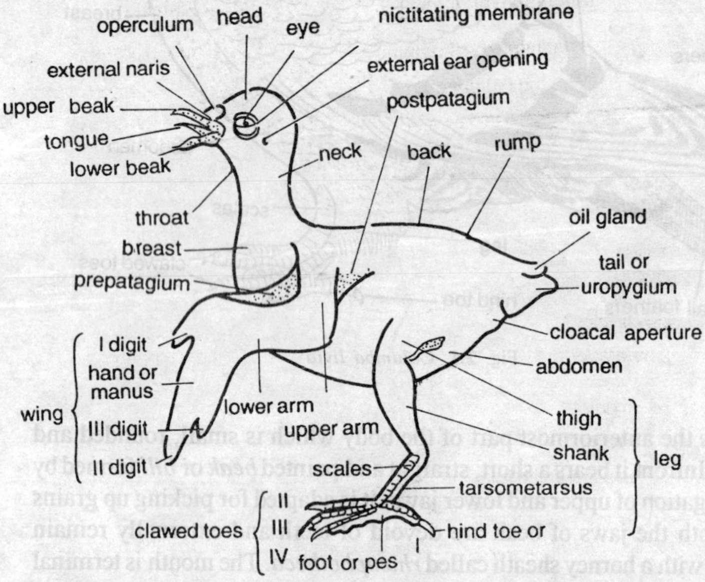

Fig. 2.2. Defeathered pigeon in lateral view.

(a) Fore-limbs

These are modified into the organs of flight i.e. wings. Each wing remains attached high on the anterio-dorsal side of trunk. Each wing consists of three parts: *upper-arm* or *brachium*, the *fore-arm* or *antebrachium* and the *hand* or *manus* with three digits or *fingers*. The digits are clawless and the second is the largest. A fold of skin called *alar membrane* or *prepatagium*, stretches between the upper-arm and fore-arm on the front side. A similar but smaller fold, *postpatagium*, is present between upper-arm and the trunk on the hind side. The elongated flight feathers arising from upper margin of the wing are called *remiges*.

When the bird is at rest the three parts of the wing are folded against the side of the trunk like the latter 'Z'.

(b) Hind-limbs

The hind-limbs or legs arise somewhat anteriorly from the trunk to balance the entire body on them. They are almost vertical. Each leg is made up of three parts-the *thigh*, the *shank* and the *foot* or *pes*. The short and muscular thigh is closely bound to the trunk and is directed downwards and forwards so that the knee is not visible. The shank is long and muscular, extends from the knee downwards and backwards. The foot is divisible into two parts, proximal slender *tarsometatarsus* and distal *digits* or *toes* which are clawed. The first digit is *hallux* directed backwards, the remaining three digits are directed forward. The feet are covered with horny dermal scales.

Tail

The tail or *uropygium* is very short and stumpy, conical projection of the trunk beyond the cloacal aperture. It is seen only in a plucked bird. It bears a fan-like group of elongated tail-feathers or *rectrices*. On its dorsal surface is a knob-like papilla, which bears the opening of a large *uropygial* or *preen gland*. This is the only integumentary gland in birds. The oily secretion of this gland is used in lubricating the feathers and also probably for keeping the horny beaks from becoming brittle.

3

Pigeon (Exoskeleton)

The exoskeleton in pigeon is purely epidermal and represented by horny claws, beak, scales, horny tooth and feathers.

Claws and Beaks

The claws of birds are typically reptilian claws over the feet. Each covers the terminal part of the digit supported internally by terminal phalanx. Unguis is a long plate of keratin sharply curved so as to enclose the smaller subunguis between its edges. At the base of claw epidermis is invaginated to from the claw root. Here, stratum germinativum continuously produces new layers of cells that become keratinized and lead to the growth of the claws. Wearing always is only at the tip.

Beaks are characterstic of birds, but are also present in certain reptiles like turtle, tortoise, mammals like duckbill and moles. Jaws are covered with hard keratin that forms the beak. Beaks or bill are regarded as large epidermal scales.

Scales

In birds epidermal scales are present on the legs and feet and on the web of aquatic birds. These are keratin derivatives. They generally overlap and are formed in the same manner as those of lizards and snakes.

FEATHERS

The feathers are characterstic of birds and birds only i.e. they do not occur in any other group of animals. They have been aptly designated the natures' masterpiece. These are light, elastic, waterproof and most important in flight. The different colours of feathers are due to the presence of melanin of various shades as well as due to iridescence. The feathers are present almost throughout the body forming the plumage. The feathers are shed annually, generally in late summer. The new ones develop from the old papillae.

Distribution

The feathers do not cover the body uniformly. They are arranged in definite tracts, the *pterylae*, separated by featherless areas, the *apteria*, which bear only a few filoplumes. The study of the arrangement of feather tracts on the body is known as *pterylosis*. The term 'plumage' refers to all the feathers collectively. The principle feather tracts or pterylae found in pigeon are as follow (its arrangement varies in different orders, families and even species of birds):

1. Cephalic or head - covering the head.
2. Cervical or neck - on the neck.
3. Humeral or shoulder - across the upper arm
4. Spinal - extending from neck to tail along the vertebral column.
5. Ventral - a double tract with one branch running along each side or the breast.
6. Wing or alar - composed of *remiges* with their *coverts.*
7. Tail or caudal - composed of *rectrices* with their coverts.
8. Femoral - spreading obliquely on outer side of the thigh.
9. Crural - covering the shank of legs.

The feathers are of serveral types, namely-quill feathers, coverts, contour, filoplume and down feathers.

Structure of a typical feather

Any contour feather from general body surface or a flight feather present on wings or tail, gives the typical structure. A typical flight feather consists of a supporting *central* or *main axis* and distal expanded upper portion, the *vane* or *vexillum.*

1. Main Axis

The axis is divided into proximal small portion, the *calamus* or the *quill* and long distal portion, the *rachis* or *shaft.*

(a) Calamus or Quill

It is a hollow, tubular and semitransparent part. At the lower end of the quill is a small opening called *inferior umbilicus.* A small, conical projection of the dermis called *dermal papilla* fits into the inferior umbilicus, through which blood vessels enter to supply nutrients and pigments to the developing feather. On the vertral side at the junction of quill and rachis, there is a second minute aperture, the *superior umbilicus.* In pigeon a small tuft of soft feathers occurs near the superior umbilicus and covers it. This tuft is known as *aftershaft* or *hyporachis.* The base of calamus remains inserted into a pit of the skin or *epidermal follicle*, from which muscle fibres pass to the feather and affect individual movement.

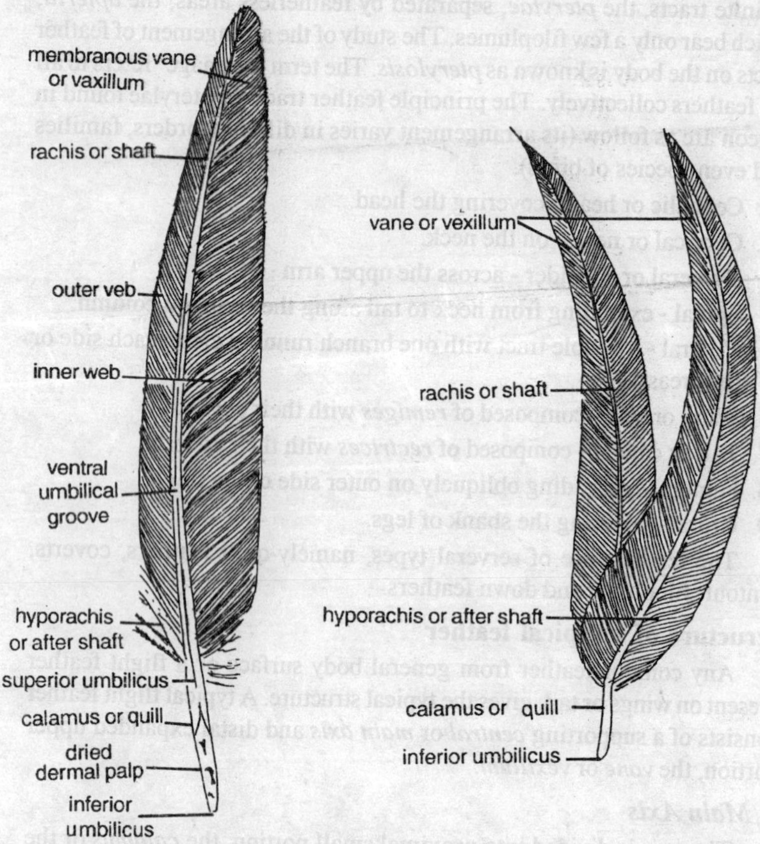

Fig. 3.1. *Columba.* A typical feather in ventral view and a quill feather of Emu.

(b) Rachis or Shaft

It forms the longitudinal axis of the vane or vexillum. It is a solid, opaque, stiff, roughly quadrangular in a section and remains filled with a closely packed mass of *pith cells.* An *umbilical groove* runs along the ventral surface of the rachis throughout its length.

2. Vane or Vexillum

It is the expanded membranous part of the feather and is divided by the rachis into two unequal lateral halves. The vane is made up by a series of numerous parallel and closely spaced, delicate thread-like structures known as *barbs* or *rami*. The barbs arise somewhat obliquely from the two lateral sides of rachis. The barbs are largest in the centre but their size gradually decreases towards the two ends of rachis. Each barb itself bears a fringe of small processes, the *barbules* or *radii* along either border. The barbules or radii are of two types-*distal barbules* directed towards the tip of feather and *proximal barbules* directed towards the base of feather. The distal barbules at the lower edge produced into minute *hooklets* or *barbicels* or *hamuli*. Similarly the proximal barbule at the upper edge bears groove or *flange*. Owing to the oblique disposition the barbules of adjacent barbs overlap and the hooklets of distal barbules becomes interlocked with the curved edges or flanges of the proximal barbules. It is the interlocking systems of barbules that gives an unbroken surface and flexiblity to the vane and also resistance against air whenever, accidently the vane surface is broken and the hooklets defasten, the birds preen or comb the barbs by means of beak or bill and set the barbules in the correct fastening position.

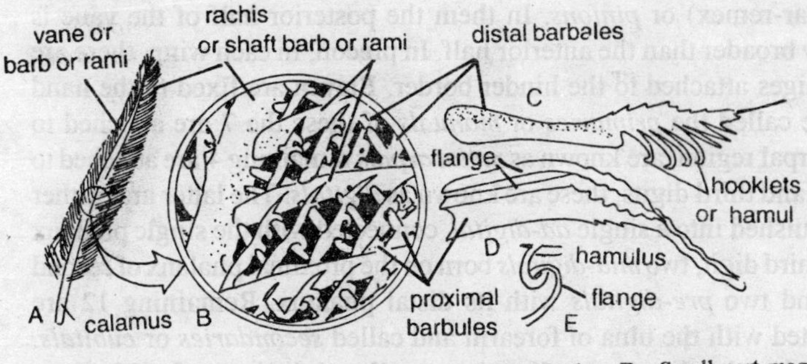

Fig. 3.2. Detailed structure of a feather. A—Complete feather. B—Small part magnified to show the interlocking mechanism. C—Distal barbule. D—Proximal barbule. E— The distal barbule interlocking with the proximal barbule.

The feathers of flightless birds like kiwi, ostrich, emu etc., lack the interlocking mechanism and are very fluffy. The aftershaft never possesses barbicles. In emu the aftershaft is almost equal to the main shaft.

KINDS OF FEATHERS

On the basis of forms, locations and functions the feathers are usually divided into several types namely, *quill feathers, coverts, contour* or *pennae, filoplumes* or *hair feathers* and *down feathers* or *plumules.*

(i) Quills or flight feathers

These are the large feathers found in the wings and tail. It has comparatively strong shaft or scapus or stem and is characterised by the presence of barbules with an interlocking arrangement.

The quills are further classified into :

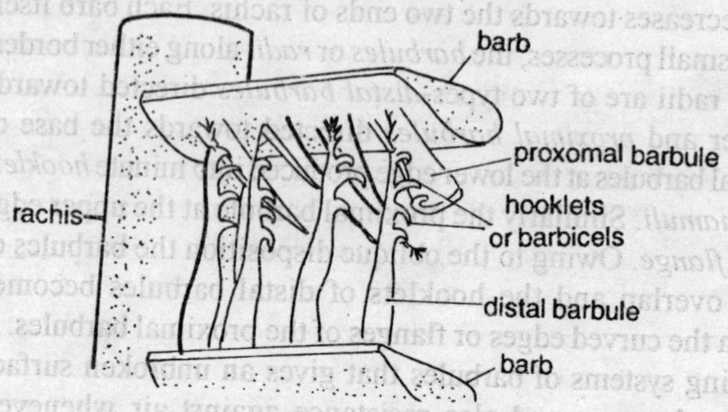

Fig. 3.3. A portion of vane magnified to show the interlocking mechainsm.

(a) Remiges

The quills found in the wings are called as *wing-quills* or *remiges* (singular-remex) or *pinions.* In them the posterior half of the vane is slightly broader than the anterior half. In pigeon, in each wing, there are 23 remiges attached to the hinder border. Eleven are fixed to the hand and are called the *primaries* or *manuals* of these the 7 are attached to metacarpal region, are known as *metacarpals* remaining 4 are attached to second and third digits, these are known as *digitals.* The latter are further distinguished into a single *ad-digital*, connected with the single phalanx of the third digit, two *mid-digitals* born on the proximal phalanx of second digit and two *pre-digitals* with its distal phalanx. Remaining 12 are connected with the ulna of forearm and called *secondaries* or *cubitals.* Attached to the thumb or pollex, is a small and distinct tuft of feathers called *ala spuria* or *bastard wing.*

(b) Rectrices

The quills found in tail are called *rectrices* or *tail quills* (singular-rectrix). In pigeon these are 12 in number and are arranged in a fan-like or semicircular manner on the tail. Here the two halves of vane are almost equal in size. These feathers help the bird as a brake and in steering the flight.

(c) Coverts

These are smaller in size and have short calamus. These are found in the wings and the tail. They fill up the gap left between the bases of quills thus presenting a continuous area to oppose the buoyancy of the air. The bases of the wing quills or remiges are covered by upper and under wing-coverts and the bases of the tail quills or rectrices by upper and under tail-coverts.

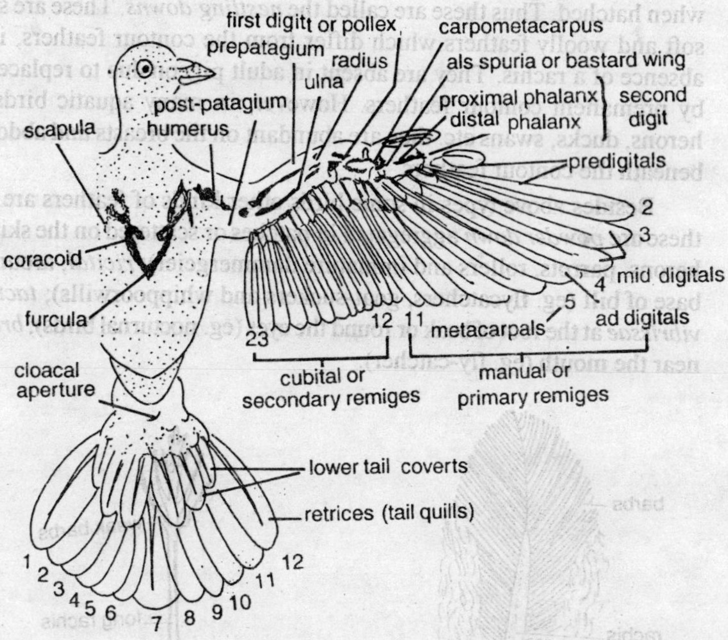

Fig. 3.4. *Columba.* Arrangement of remiges and rectrices.

(ii) Contour feathers or pennae

As the name implies, these feathers form the general covering of the body and give the animal a definit shape or contour. Structurally these feathers consisting of a central axis and vane. The barbs are not so strongly joined and can be separated easily.

Contour feathers provide a smooth surface to the body, thus reducing frictional resistance to motion through air. Moreover, they make a good heat-conserving surface.

(iii) Filoplumes

They are also known as *hair feathers* or *pin feathers*. They occur beneath the contour feathers and visible in a plucked animal only. They are extremly small in size consisting of a short calamus, and a long thread like rachis with a few barbs and barbules at the free tip.

Their function is unknown.

(iv) Down feathers

The down feathers form the natal covering of the newly hatched birds providing excellent insulation. Although many birds are quite nacked when hatched. Thus these are called the *nestling downs*. These are small, soft and woolly feathers which differ from the contour feathers, in the absence of a rachis. They are absent in adult pigeon due to replacement by premanent contour feathers. However, in many aquatic birds like herons, ducks, swans etc. they are abundant on the breasts and abdomen, beneath the contour feathers.

Besides above types in some birds other kinds of feathers are seen, these are *powder down* aggregated in patches or scattered on the skin (eg. herons, parrots, rollers and immature dammergeier); *ricital*, around the base of bill (eg. flycatchers, goat-suckers and whippoorwills); *tactile* or *vibrissae* at the root of beak or round the eyes (eg. nocturnal birds); *bristles*-near the mouth (eg. fly-catcher).

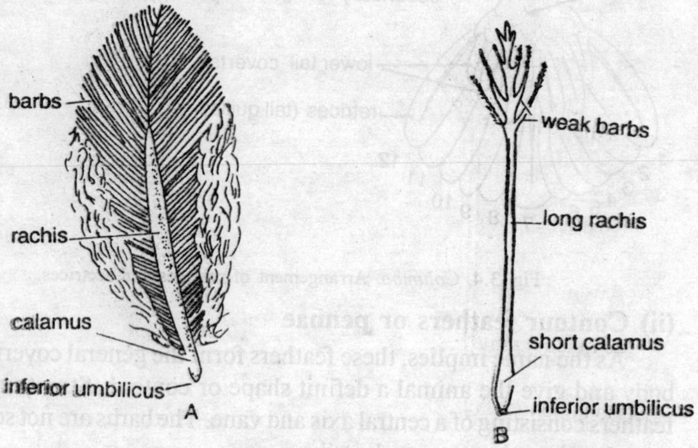

Fig. 3.5. Feathers of a pigeon. A—Contour feather. B—Filoplume.

Colouration

Feather colour variation results from either one or both of two mechanisms physical structure and chemical pigments. White, blue, and irridescent colours are structurally derived, where as red, yellow, brown, and black are of pigment origin. Brown and black-coloured feathers are caused by melanin taken up from a reservoir of melanocytes in the germinal region of the follicle. There is no pigment in white feather, the presence of air in cells and the polygonal shape of the barbule cells simply break up the light and reflect and refract the several wave lengths equally. Structural blue results from the physical scattering action of minute pores or particles in the cells located beneath the outer layer of feather barbs. Blue wave lengths are scattered, the longer ones being transmitted, because of this scattering, such areas appear blue in reflected light. Many birds exhibit strikingly complicated variations of colour patterns produced by contributions from the overlapping of contour feathers.

DEVELOPMENT OF FEATHER

A feather originates as a small projection on the skin in the embryo like a reptilian scale or mammalian hair.

Development of Down feathers

On the 5th or 6th day of incubation small projections are seen on the delicate and still semitransparent skin of the embryo. These are feather *papillae* or *dermal papillae*. Within each papilla dermis extends and maintains a supply of materials necessary for the development of a feather. The papilla soon grows into a conical cylinder, at the same time, it sinks into a depression, the *annular groove*, around the base of papilla. At the same time, it sinks into a tubular pit, the *feather-follicle*. Papilla has a core of dermal palp in the centre surrounded by the stratum germinativum

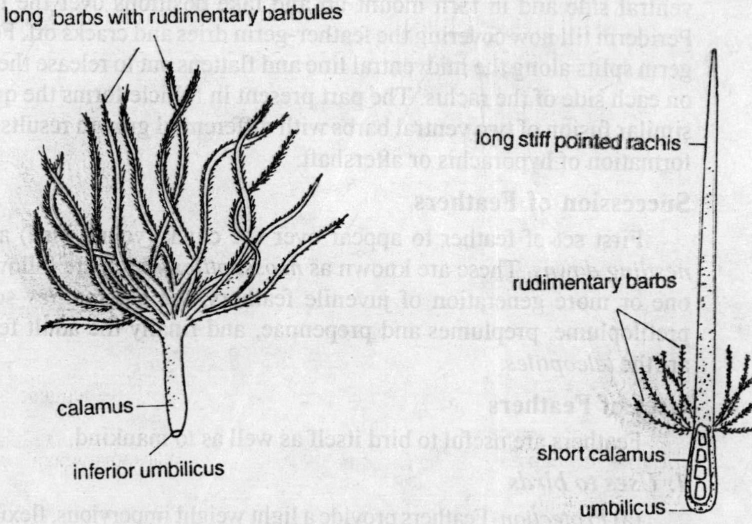

Fig. 3.6. A nestling down feather of a young pigeon and rictal bristle.

and the cornified cells of stratum corneum. The cornified layer is known as *periderm* and the developing feathers is a *feather germ* at this time. Actively dividing cells of the stratum germinativum become arranged into longitudinal columns that will change into barbs. A feather-germ with rudiments of barbs is called a *pin-feather* and projects well above the skin. Activity in the stratum germinativum gradually comes to a stand still as the dermal palp becomes progressively less nutritive. The periderm splits from the top and the barbs spread out. Parts of pin-feather within the feather follicle becomes *quill* or *calamus* maintaining a connection with the papilla below by the *inferior umbilicus*. Dermal palp within the quill dries into pith.

Development of Contour feather

In case of contour feather, a fresh papilla forms at a deeper level, so that the down feather drops off from the apex of the latter. It develops the same way up to the stage when longitudinal ridges of stratum germinativum have been formed, after this the development is somewhat more complex. Unequal growth on one side, more rapid presumably on the dorsal side then on the ventral side causes two adjacent longitudinal ridges to outgrow other and unite all along their length to from the rachis. The line of union being represented in the fully developed feather by the umbilical groove. As the rachis elongates, other ridges successively mount up and take positions over the rachis, new ridges or barbs appear on the ventral side and in turn mount up and take positions over the rachis. Periderm till now covering the feather-germ dries and cracks off. Feather germ splits along the midventral line and flattens out to release the barbs on each side of the rachis. The part present in follicle forms the quill. A similar fusion of two ventral barbs with differential growth results in the formation of hyporachis or aftershaft.

Succession of Feathers

First set of feather to appear over the chick (young bird) are the *nestling downs*. These are known as *neossoptiles* which are followed by one or more generation of juvenile feathers, the *messoptiles* such as prefiloplume, preplumes and prepennae, and finally the adult feathers are the *teleoptiles*.

Uses of Feathers

Feathers are useful to bird itself as well as to mankind.

1. Uses to birds

(a) *Protection*. Feathers provide a light weight impervious, flexible but resistant and durable body cover. They protect the skin from external injuries.

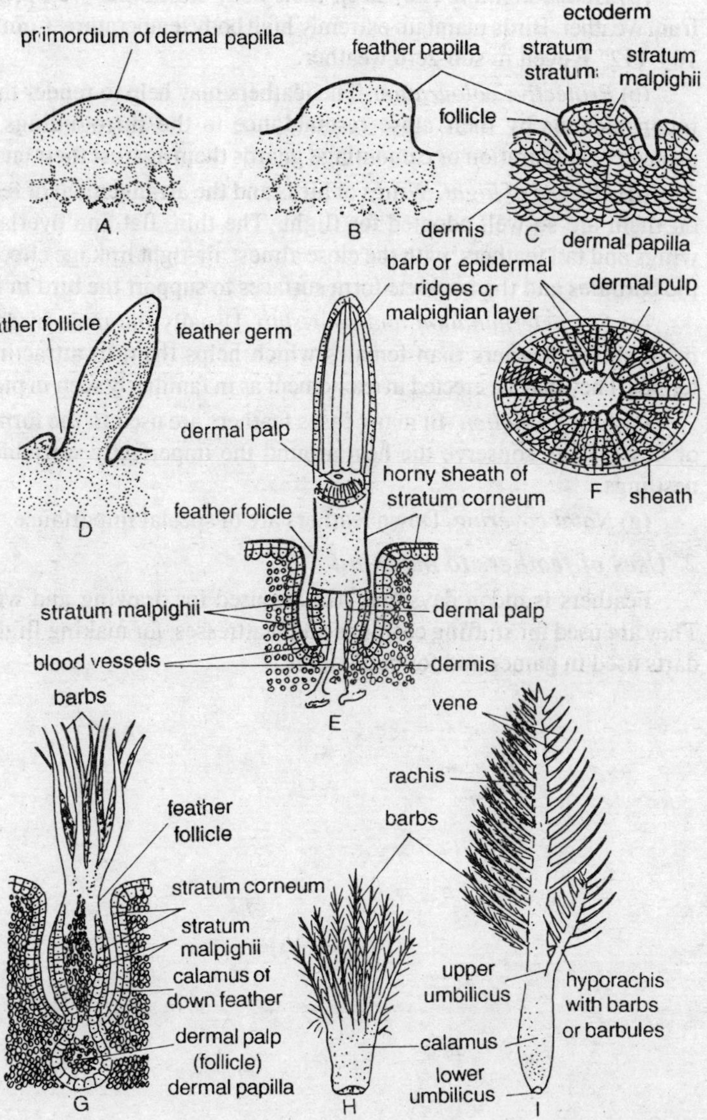

Fig. 3.7. Stages in the development of a feather. A—Beginning papilla. B—Feather papilla arising above skin surface. C—Same in V.S. D—Feather germ. E—Feather germ in V.S. F—Feather germ in T.S. G—Down feather in its follicle. H—Mature down feather newly hatched bird. I—Young contour feather.

(b) *Heat retention.*They keep their body warm and well protected from weather. Birds maintain extremly high body temperature. Commonly 104-112 °F even in sub-zero weather.

(c) *Protective colouration.* The feathers may help to render the bird inconspicuous by their close resemblance to the surroundings. This protective colouration or camouflage guards them against their enemies.

(d) *Organs of flight.* Wings of birds and the arrangement of feathers on them are so well adopted for flight. The thin, flat and overlapping wings and tail feathers, with the close almost air-tight linkage effected by the barbules and the hooklets form surfaces to support the bird in flight.

(e) *Sex determination and courtship.* Usually the male birds have more showy feathers than females which helps them in attracting the female.They may be erected in excitement as in familiarly seen in pigeons.

(f) *Nest formation.* In many cases feathers are used in the formation of nests. They conserve the heat around the imperfectly warmblooded nestlings.

(g) *Natal covering.* Down feathers are of special importance.

2. Uses of feathers to mankind

Feathers is olden days and still are used for drawing and writing. They are used for stuffing cushions and mattresses, for making flights for darts used in games and for battledores.

4

Endoskeleton of fowl

Skeleton of dermal origin forms the framework of the body and is called the *endoskeleton* in contrast to *exoskeleton* derived from the skin and lying either above or within the limits of skin. Skeletal tissue such as bones and cartilages together with connective tissue are the structural elements of the endoskeleton. The endoskeleton of fowl is fully ossified and lightly built on the 'hollow girder principle'. The large bones are usually *pneumatic* i.e.*spongy* or *hollow* being filled with air spaces in continuation with the air spaces from the lungs.

The endoskeleton comprises the usual two divisions :

The *axial skeleton* including skull, vertebral column, sternum and ribs, and the *appendicular skeleton,* including bones of limbs and girdles.

I. AXIAL SKELETON

Skull

It is characterised by very light weight, fusion of most of the bones so that it has many sutures, very large orbits, *monocondylic* condition (single occipital condyle) and tooth less conical beak in front. The skull is *tropibasic* i.e the cranium does not extend forward into orbital region. The jaw suspension is *autostylic*. The skull palate is *schizognathous* i.e. with short vomers allowing the palatines to meet.

The skull can be studied under three heads :

(1) Cranium, (2) Sense capsules and (3) Visceral skeleton that contains jaws and hyoid apparatus.

1. Cranium

The cranium forms the posterior median hollow part of the skull housing the brain during life. It has many perforations for the passage of cranial nerves emerging from brain and for blood vessels. It is further divided into following regions:

(I) Posterior, *occipital region*, (ii) Middle, *parietal region* and (iii) Anterior, *frontal region*.

(i) Occipital region

It is the posterior most part of skull and is formed by the usual four bones, the *basioccipital* which forms the base of cranium, two lateral *exoccipitals* and the *supraoccipital* which forms the roof of cranium. All these bones enclose a large rounded opening called *foramen magnum*. The spinal cord passes through this foramen to join the brain. Just below the foramen magnum there is a single rounded knob like structure called *occipital condyle*, which is formed by basioccipital bone. In frog and rabbit there are two lateral occipital condyles which are formed by exoccipitals. A prominent semicircular ridge of bone called *lambdoidal crest* is present where supraoccipital joins the parietal.

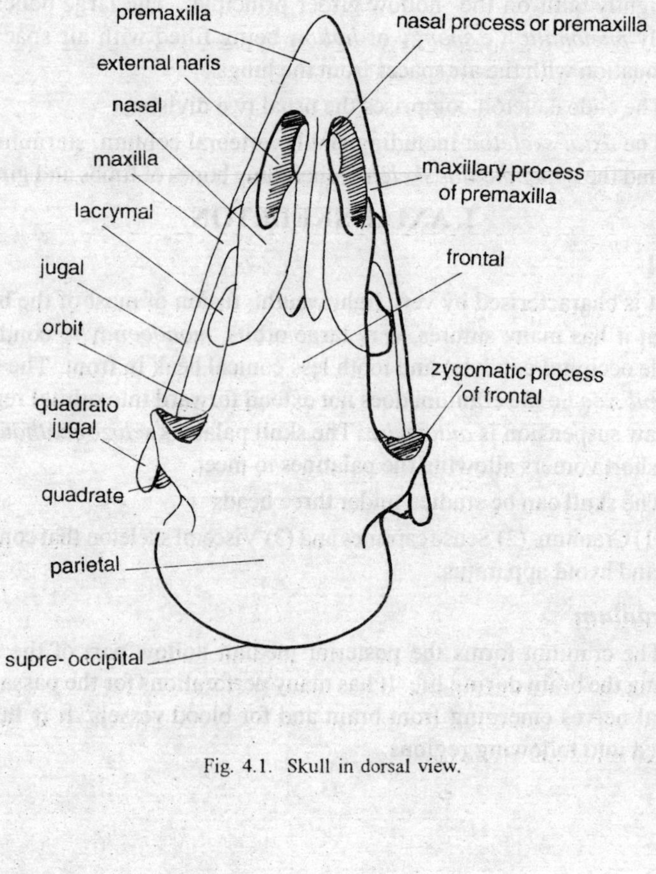

Fig. 4.1. Skull in dorsal view.

(ii) Parietal region

It lies infront of occipital region. It is made up of five bones, a pair of *parietals*, a pair of *squamosals* and a *basitemporal*. The roof of parietal region is made up of fused parietals which are squarish and are present in front of lambdoidal crest. The lateral sides are formed by the *squamosals*. Each squamosal is produced into a slender *zygomatic process*. The zygomatic process extends forward and slightly downwards often meeting and fusing distally with lower end of zygomatic process of frontals. The basitemporal is a broad, transverse bone on the ventral side in front of basioccipitals. It covers the basisphenoid. A small anterior triangular part of the basisphenoid remains uncovered. This part has basipterygoid process for the pterygoids. The basisphenoid is continued forwards by a slender *parasphenoid rostrum*.

(iii) Frontal region

It is the anterior most part of cranium, lying infront of the parietal

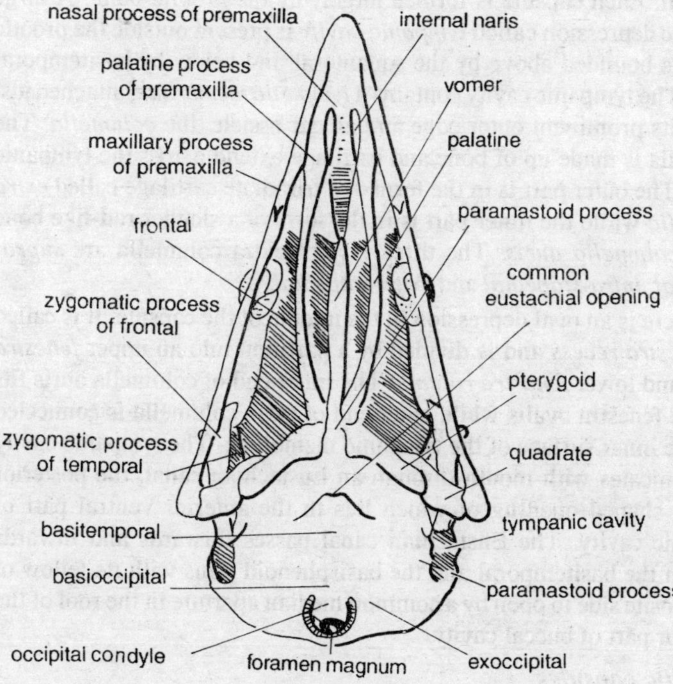

Fig. 4.2. Skull in ventral view.

region. It is made up of five bones, a pair of *frontals*, a *rostrum* and a pair of *alisphenoids*. The frontals are long bones forming the roof of frontal region. The lateral posterior end of each frontal is drawn out into *zygomatic process* or *post-orbital process* which joins the zygomatic process of squamosal. The alisphenoids are broad, semivertical bones continued forward into the *orbito-sphenoids* forming the sides of the frontal region. The rostrum is a slender, median rod-like bone in front of the basitemporal bone. It forms the lower thickened border of the *interorbital septum*.

2. Sensory capsules

The sensory capsules lodge and protect the special sensory organs. These remain closely attached to the cranium. The sensory capsules include paired (i) Auditory capsules (ii) Optic capsules and (iii) Olfactory capsules.

(i) Auditory capsules

These are enclosing the organs of hearing i.e. the internal ears. They are attached behind, one on either lateral side of the occipital region of the skull. Each capsule is formed largely by the *prootic bone*. A large cup- like depression called *tympanic cavity* is present outside the prootic which is bounded above by the squamosal and below by basitemporal bones. The tympanic cavity contains a *tympanic membrane*, attached just within its prominent outer edge and an ear ossicle, the *columella*. The columella is made up of bone and cartilage extend across the tympanic cavity. The outer part is in the form of a triradiate cartilage called *extra columella* while the inner part is in the form of a slender rod-like bone called *columella auris.* The three rays of extra columella are *supra-stapedial*, *infra-stapedial* and *extra-stapedial.*

There is an oval depression in the middle of the capsule. It is called the *fenestra recess* and is divided by a partition into an upper *fenestra ovalis* and lower *fenastra rotunda* The inner end of columella auris fits into the fenestra ovalis while outer end of extra columella is connected with the inner surface of the tympanic membrane. The tympanic cavity communicates with mouth through an Eustachian canal, the posterior funnel- shaped opening of which lies in the anterior ventral part of tympanic cavity. The Eustachian canal passes forwards and inwards between the basitemporal and the basisphenoid joins with its fellow of the opposite side to open by a common median aperture in the roof of the posterior part of buccal cavity.

(ii) Optic capsules

The optic capsules or orbits are very large to accommodate large eyes. They are separated from one another by a thin, longitudinal, vertical

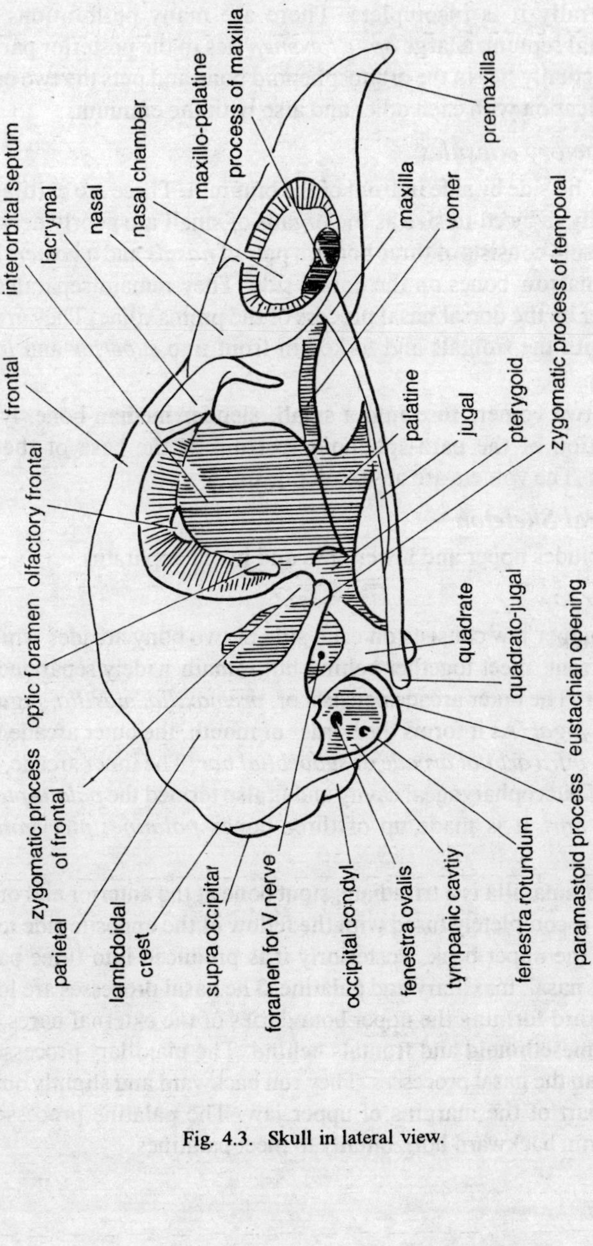

Fig. 4.3. Skull in lateral view.

partition called *interorbital septum*. This septum is formed by the fusion of the *presphenoid* and *orbitosphenoid* in the posterior part and by *mesethmoid* in the front and upper part. Each orbit is bounded anteriorly by the *frontals* and *lacrymal*, dorsally by the *frontals* and *alisphenoid* and ventrally it is incomplete. There are many perforations in the interorbital septum, a large *optic foramen* lies in the posterior part of the orbit. It actually lies in the orbitosphenoid bone and puts the two orbits in communication with each other and also with the cranium.

(iii) Olfactory capsules

They lie side by side in front of the cranium. These are cartilaginous and greatly reduced in size as the organs of smell are poorly developed. Each capsule consists of three bones a pair of *nasals* and a *vomer*. Nasals are long narrow bones on the dorsal side. They remain separated from each other by the dorsal nasal process of the premaxillae. They are fused behind with the frontals and forked in front into *superior* and *inferior processes*.

The two vomers fuse into a small, slender median bone, lying in continuation of the para-sphenoid rostrum at the base of the nasal chambers. The vomers are absent in pigeon.

3. Visceral Skeleton

It includes upper and lower jaws and hyoid apparatus.

(i) Upper jaw

The upper jaw consists on each side of two bony arcades which are fused in front, meet together behind but remain widely separated from each other. The outer arcade consists of: *premaxilla, maxilla, jugal,* and *quadrato-jugal*. As it forms the border of mouth, the outer arcade is also termed as *infra orbital arcade* or *suborbital bar*. The inner arcade, forms the roof of buccopharyngeal cavity and is also termed the *palato-pterygo-quadrate bar*. It is made up of three bones *palatine*, *pterygoid* and *quadrate*.

The premaxilla is a triradiate, stout bone at the anterior end of outer arcade. It is completely fused with the fellow of the opposite side to from the tip of the upper beak. Posteriorly it is produced into three pairs of processes, nasal, maxillary and palatine. The nasal processes are longest run backward forming the upper boundaries of the external nares. They join with mesethmoid and frontals behind. The maxillary processes are shorter than the nasal processes. They run backward and slightly outward forming part of the margins of upper jaw. The palatine processes are smallest, run backward horizontally to meet palatines.

The maxillae are slender, rod like bones. Each maxilla lies behind and to the inner side of maxillary process of the premaxilla. Anteriorly it is produced inward into a thin spongy horizontal lamina called, *maxillopalatine process*. With these processes the palatines are attached on ventral side. It forms the anterior half of the suborbital bar.

The quadrato-jugal is a slender, bony rod behind the maxilla forming the posterior half of the suborbital bar. It articulates behind with the outer side of distal end of the quadrate bone.

The jugal is slender rod like bone forming the middle part of the upper border of the suborbital bar lying dorsally upon the other two components of bar viz. the maxilla and quadrato-jugal.

The quadrate is triradiate stout bone lying just in front of the tympanic cavity and forming the suspensorium with the squamosal bone of the skull. The *dorsal arm* movably articulates with the squamosal between the base or its zygomatic process and the tympanic cavity. The ventral arm forms a *condyle* for articulation with the mandibles. Its *anterior arm* or *orbital process* runs inwards parallel to the pterygoid.

The palatine is a long, slender slightly curved bone lying infront of pterygoid. It meets the palatine process of premaxilla in front and pterygoid behind. Posteriorly it is expended into a broad lamella, the sides of which articulate with the rostrum.

The pterygoid is a short but stout and rod-shaped, obliquely placed bone forming the posterior roof of the mouth cavity. It articulates internally with the palatine and rostrum and externally with inner surface of the quadrate.

(ii) Lower jaw or mandible

The lower jaw appears to be a single toothless 'V'shaped bone but actually it is composed of a pair of long, curved and laterally compressed halves or rami, which are fused in front of the middle line. Each ramus is

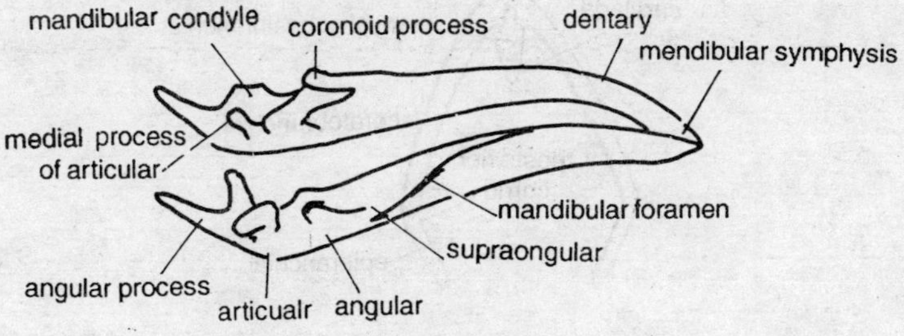

Fig. 4.4. Lower jaw.

made up of five bones, the *articular, angular, supra-angular, dentary* and *splenial*. The first one is a replacing bone while rest of the four bones are investing bones.

The articular forms the expanded posterior end of each ramus. On the dorsal surface it bears an elongated mandibular condyle for the articulation with the quadrate. Posteriorly it is produced into medial and angular processes. The angular is a splint-like bone along the ventral and inner surface of the posterior half of the ramus. The supra-angular bearing a small *coronoid process* which forms the upper border of posterior third of the ramus. The dentary is a large bone and forms about the complete distal half of the ramus. In front it fuses firmly with its fellow of the opposite side at a *symphysis*. Posteriorly it extends between the angular and the supra-angular bones up to the articular. A *mandibular foramen* is present. The splenial is a thin plate on the inner surface of the middle part of the ramus.

(iii) Hyoid apparatus

The hyoid apparatus lies in the floor of the buccal cavity, supporting the tongue. It is made up of bones and cartilages. The median arrow-shaped body is composed of three bones an anterior *entoglossal* or *ceratohyal* bearing a cartilaginous process infront, a mediam *basihyal* and a posterior *urohyal* or *basibranchial* having a cartilage behind. Two pairs of cornua are present, a pair of short *anterior cornua* project *backwards* from the entoglossal, representing the ceratohyals. A pair of long, backwardly directed *posterior cornua* or *thyrohyals* arising from

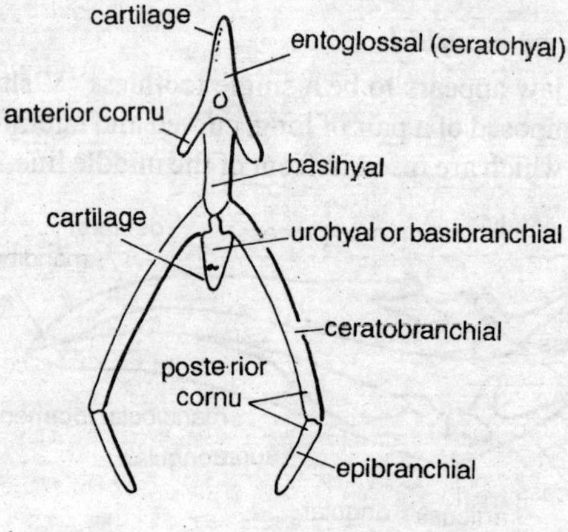

Fig. 4.5. Hypoid apparatus.

the sides of the body between the basihyal and urohyal. These represent the ceratobranchials and epibranchials and are connected by cartilage.

Vertebral Column

The vertebral column consists of about 40 vertebrae. It is divisible into four regions - a long and flexible *cervical region* containing 16 vertebrae, a rigid *thoracic region* containing 7 vertebrae, a rigid *sacral region* containing 16 vertebrae, a *caudal region* containing 4 and 5 vertebrae and a terminal *pygostyle*. The ends of the centra of vertebrae are *heterocoelous* or saddle-shaped.

1. Cervical vertebrae

The neck is long so that the cervical region is also long, The first cervical is called the *atlas*, the second cervical is called *axis*, from 3rd to 10th which are almost similar are called *typical cervical*. Behind 10th the cervical are modified hence called *posterior cervical*.

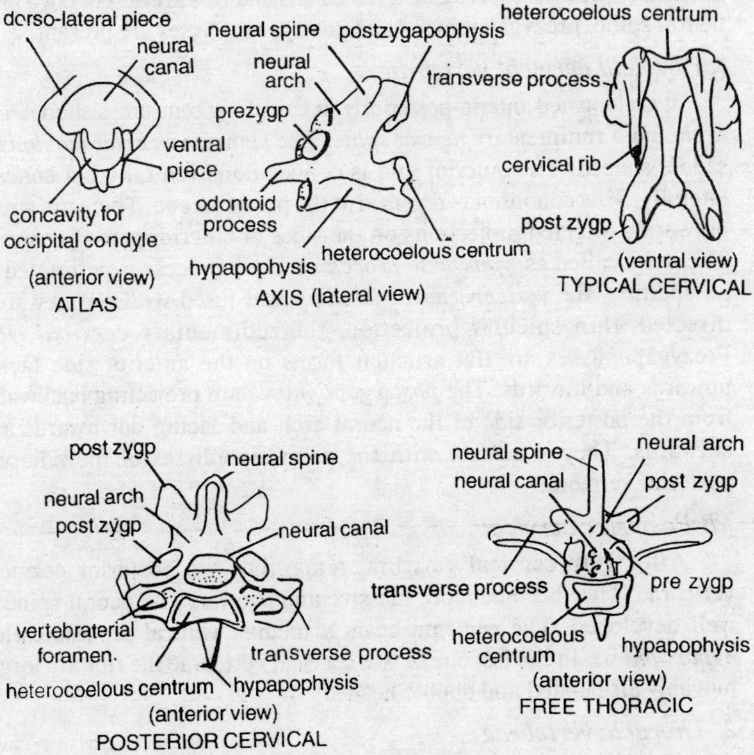

Fig. 4.6. The cervical and thoracic vertebrae.

(a) Atlas or first cervical vertebra

It is the first vertebra and is ring-like in appearance. Centrum is reduced, neural spine, transverse process and pre-zygapophyses are absent. The post-zygopophyses are very small. It consists of three elements, a single ventral and two dorsolateral joining mid-dorsally to form the neural arch. A cup like articular surface is found at the anterior end of the centrum to receive the single occipital condyle of the skull. In life, its cavity or neural canal is divided by a transverse ligament into an upper spinal canal for spinal cord and a lower odontoid canal for the odontoid process of axis.

(b) Axis or second cervical vertebra

It is also called *epistropheus* and is bigger than the atlas but smaller than rest of the cervicals. The centrum is produced infront into a peg-like *odontoid process* which fits into the odontoid canal of atlas vertebra. The transverse processes, vertebrarterial canals and ribs are absent but a blunt neural spine, prezygapophyses and postzygapophyses are present.

(c) Typical cervical vertebrae

It is elongated anterio-posteriorly. It has a long centrum, a short *neural arch* and a rudimentary *neural spine*. The centrum is *heterocoelous* or saddle shaped. The anterior end is convex dorsoventrally but concave laterally. This condition is reversed at the posterior end. There are small, outwardly directed projections on the sides of anterior part of centrum. These are called as *transverse processes*. Each process is perforated by an aperture, the *vertebrarterial foramen* and fused with a backwardly directed, thin spicular projection, the rudimentary *cervical ribs*. Prezygapophyses are flat articular facets on the anterior side facing upwards and inwards. The *postzygopophyses* are projecting backwards from the posterior side of the neural arch and facing downwards and outwards. They articulate with the prezygapophyses of the adjacent posterior vertebra.

(d) Posterior cervical

After 10th cervical vertebra, remaining are posterior cervical vertebrae. They become more massive and shorter. The neural spine is well developed. The centrum bears a median ventral process called *hypapophysis*. In the last one or two cervical vertebrae the ribs are large, movably articulated and double headed.

2. Thoracic vertebrae

There are seven thoracic vertebrae. The 2nd to 5th are fused together forming *fused thoracic vertebrae* while Ist and 6th they are free and are called *free thoracic vertebra*. The 7th vertebrae is fused with the synsacrum.

(a) Fused thoracic vertebrae

Four vertebrae i.e, second, third, fourth and fifth are fused together in such a way that their transverse processes forming horizontal bony

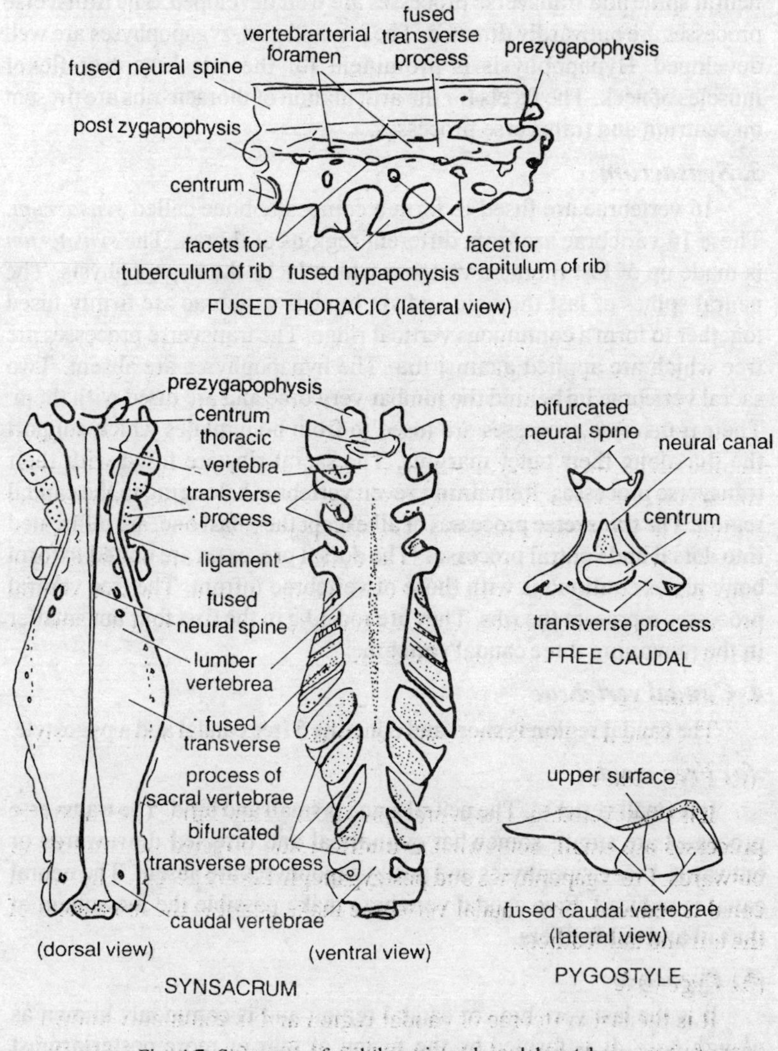

Fig. 4.7. Synsacrum, fused thoracic and caudal vertebrae.

plates. Their neural spines forming a continuous mid-dorsal ridge. Their hypapophyses forming a similar but perforated mid-ventral ridge. The fused centra and transverse processes bear on either side facets for articulation of thoracic ribs.

(b) Free thoracic vertebrae

The centrum is well developed and is heterocoelous. Neural arch, neural spine and transverse processes are well developed. The transverse processes are outwardly directed. The pre-and post-zygapophyses are well developed. Hypapophysis is prominent for the attachment of flexor muscles of neck. The facets for the articulation of thoracic ribs are present on centrum and transverse processes.

3. Synsacrum

16 vertebrae are fused to form a composite bone called *synsacrum*. These 16 vertebrae are from different regions of thorax. The *synsacrum* is made up of last thoracic vertebra which lacks the hypapophysis. The neural spines of last thoracic and six lumbar vertebrae are firmly fused together to form a continuous vertical ridge. The transverse processes are free which are applied against ilia. The hypapophyses are absent. Two sacral vertebrae lie behind the lumbar vertebrae and are fused with them. Their transverse processes are fused to from bony plates which support the ilia along their outer margins. The sacral ribs are fused with their transverse processes. Remaining seven vertebrae belonging to the caudal region. The transverse processes of all except the—last one, are bifurcated into dorsal and ventral processes. The dorsal processes are united to form bony plates continuous with those of vertebrae infront. The free ventral processes represent the ribs. They are rod-like in the first four but smaller in the remaining three caudal vertebrae.

4. Caudal vertebrae

The caudal region is short and contains 5 free caudal and a pygostyle.

(a) Free caudal

It is small vertebra. The neural spine is small and bifid. The transverse processes are small, somewhat cylindrical and directed downwards or outwards. Prezygapophyses and postzygapophyses are absent. The neural canal is reduced. Free caudal vertebrae make possible the movement of the tail and tail feathers.

(b) Pygostyle

It is the last vertebrae of caudal region and is commonly known as *ploughshare*. It is formed by the fusion of four or more posteriormost caudal vertebrae. It is laterally compressed with a straight dorsal border and a convex ventral border. The pygostyle provides attachment to caudal muscles and supports the large tail feathers.

Sternum

It is also known as *breast bone* because it lies in the breast region. It is very large in size and is situated in the thoracic and abdominal regions and it supports them. Its long median part is called the *body of sternum*. The body of sternum is wide and stout infront, narrow in middle and widens again behind. It is concave on upper side bearing a vertical crest or *keel*. The keel or carina is a triangular stout bony plate to which are attached the flight muscles. Anteriorly, a manu-brium is present which is a small vertical plate having an aperture. A pair of large processes arise from the lateral sides of sternum and extend backwards. These are called *metacostea* or *xiphoid processes*. Each process is divided into two large processes at its base. These are called *oblique lateral xiphoid process* and *posterior lateral xiphoid process*. In front of metacostea, the body of sternum bears notches on its lateral border for the sternal part of ribs. These borders are known as *costal surfaces.* Transverse grooves, called *coracoid grooves* are present on the base of the coracoid bones.

Between the coracoid grooves and the costal surfaces the body of the sternum sends upwards a *costal process* also called *anterior metasternal process.*

Ribs

There are seven pairs of ribs in fowl, two pairs of cervical ribs and five pairs of thoracic ribs.

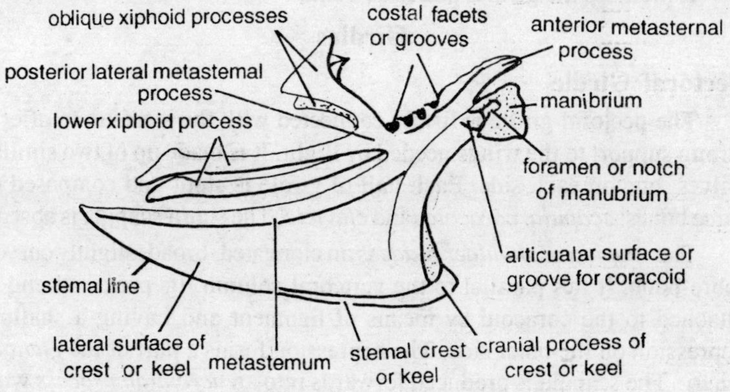

Fig. 4.8. Sternum.

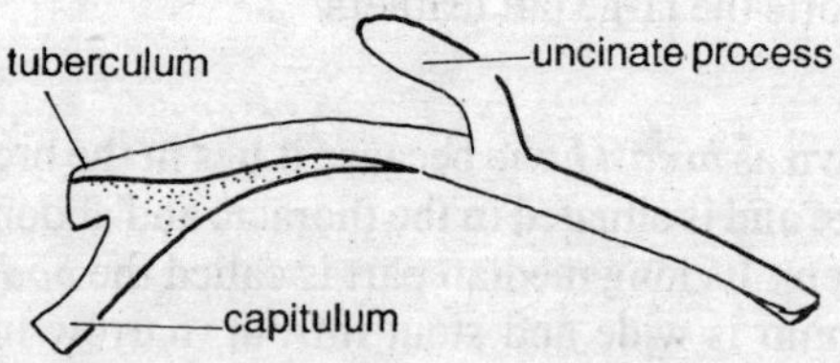

Fig. 4.9. Thoracic rib.

(a) Cervical ribs

There are two pairs of cervical ribs with movably joints. The first is small but has a distinct *capitular* and *tubercular facets*. The second is large and bears an *uncinate process* which rests on succeeding ribs.

(b) Thoracic ribs

These are completely bony and increase in size from before backwards. Each rib is made of two parts a dorsal *vertebral part* and a ventral *sternal part*. The proximal end of vertebral part of each rib is double headed consisting of lower *capitulum* and upper *tuberculum*. Capitulum attached to the transverse process of the thoracic vertebrae. An uncinate process is found in the posterior region of vertebral part of all the thoracic ribs except the last one.

II. APPENDICULAR SKELETON

It includes the girdles and limb bones.

Girdles

Pectoral Girdle

The pectoral girdle is firmly connected with the sternum to offer a strong support to the wings needed by flight. It is made up of two similar halves, one on each side. Each half of girdle is stout and composed of three bones: *scapula, coracoid,* and *clavicle*. The supra scapula is absent.

The *scapula* or *shoulder blade* is an elongated, broad, slightly curved sabre-bone. It lies parallel to the vertebral column. Its proximal end is attached to the coracoid by means of ligament and having a shallow depression on the outer side. The depression forms a part of the *glenoid cavity.* The scapula is produced forwards into an *acromian process* with which the clavicle is attached.

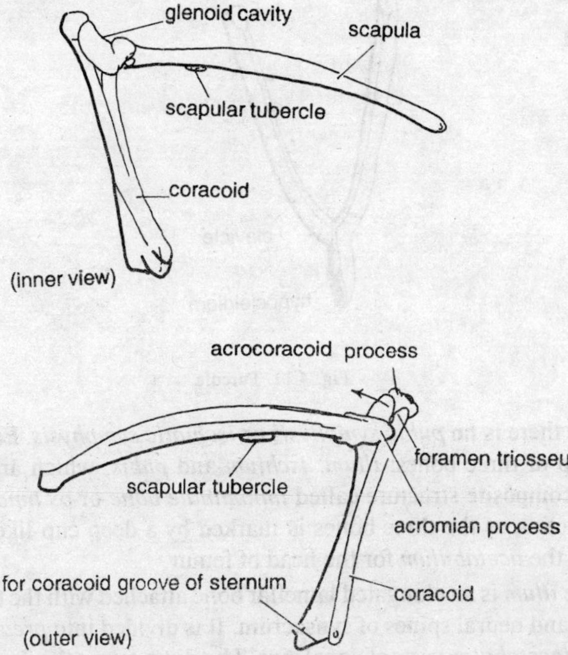

Fig. 4.10. Pectoral girdle.

The *coracoid* is a stout, long and slender bone. Its proximal end is attached with the anterior edge of the sternum. while the distal end extends forwards, upwards and somewhat outwards. The distal end gives out a process, the, *acrocoracoid process* which meets clavicle. The epicoracoid and procoracoids are absent. The distal end bears on its outer side a shallow depression, which forms the remaining part of glenoid cavity. The head of humerus lodges in *glenoid cavity*.

The *clavicles* are slender, slightly curved and delicate rod-like bones. The upper end of each clavicle is attached partly to the acrocoracoid process of coracoid and partly to the acromian process of the scapula to enclose a circular gap called the *foramen triosseum*. The lower ends of clavicles are fused to a small interclavicle to form a median laterally compressed disc or *hypocleidium*. Two clavicles and inter-clavicle or hypocleidium together form a V-shaped structure known as *furcula* or *wish-bone* or *merry-thought*.

Pelvic Girdle

The pelvic girdle is large and firmly united with the synsacrum along whole of its length. It consists of two similar halves, which remain separate

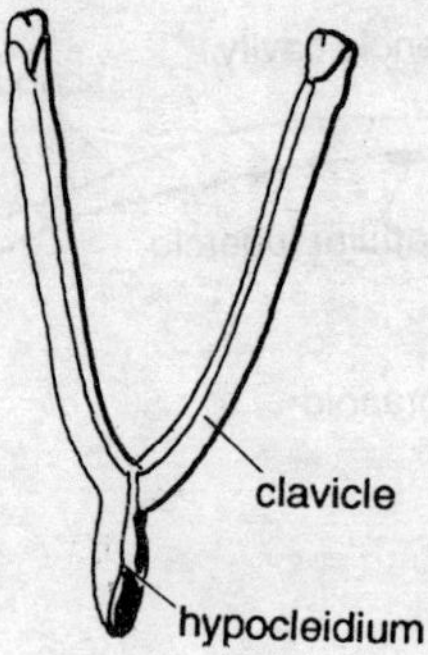

Fig. 4.11. Furcula.

because there is no *pubic symphysis* or *ischiatic symphysis*. Each half is made up of three bones; *ilium, ischium* and *pubis*, which are fused to form a composite structure called *innominate bone* or *os innominatum*. The junction of the three bones is marked by a deep cup-like articular surface, the *acetabulum* for the head of femur.

The *ilium* is an elongated lamellar bone attached with the transverse process and neural spines of synsacrum. It is divided into *preacetabular* and *postacetabular* parts of equal size. The ilium forms the dorsal part of the acetabulum. It bears just above acetabulum a small projection, the *anti-trochanter* for articulation with the great trochanter of femur. It has two concavities on the undersurface for kidney lobes.

The *ischium* is a broad, lamellar, nearly vertical bone behind the acetabulum. It is fused posteriorly with the post-acetabular part of ilium but separated anteriorly from it by a large oval *ilio-ischiatic foramen*.

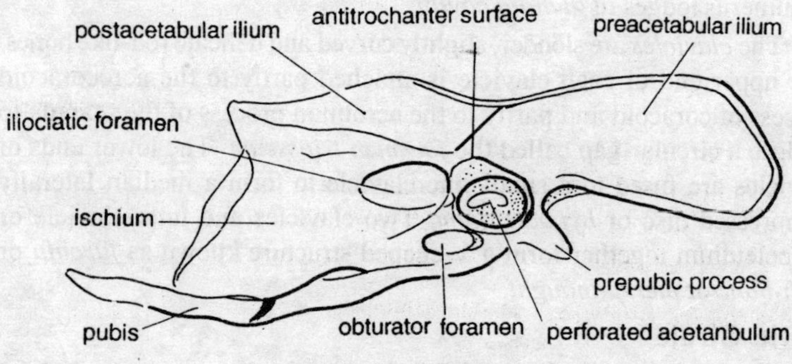

Fig. 4.12. Pelvic girdle.

The *pubis* is long, thin and curved bone running along the ventral edge of the ischium. It is separated from ischium along its greater length by a small, oval *obturator foramen* or *ischio-pubic foramen*. The pubis extends in front of the acetabulum as *preacetabular* or *prepubic process*. The free posterior end of pubis extends backwards for some distance beyond the ischium.

LIMB BONES

Limb bones are grouped into two categories:

I. Bones of Forelimbs

There are three bones in the each forelimb. The upper arm contains a single bone, the *humerus*, the fore- arm contains two bones *radius* and *ulna* and the *hand* or *manus* has many bones.

1. Humerus

It is a stout, elongated and slightly curved bone, expanded at both the ends. It is differentiated into three parts, proximal end having smooth, oval and knot-like *head*, middle slightly curved long *shaft* and distal end having a pulley-like *trochlea*. Head articulates with the *glenoid cavity* of pectoral girdle. It is bordered by two projections, a small *lesser tubercle* or *tuberosity* and a *greater tubercle*. The lesser tubercle shows a prominent *deltoid ridge* for the insertion of flight muscles. The greater tubercle possesses a large *pneumatic foramen* communicating with the air cavity of shaft. The trochlea bears two knob like articular surfaces, a *preaxial* for the radius and *postaxial* for the ulna.

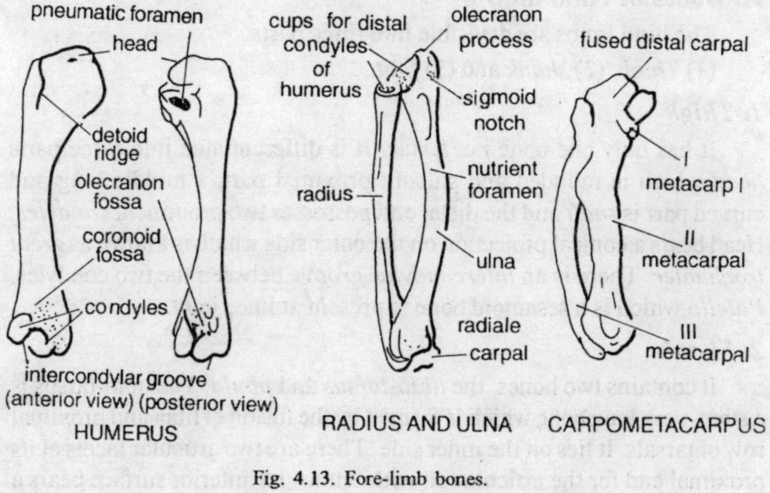

Fig. 4.13. Fore-limb bones.

2. Radius-ulna

These are two bones. The *radius* is shorter, slender, straight and preaxial. Its proximal end has a cup-like articular surface for the preaxial condyle of the trochlea of humerus. Distally it articulates with carpals.

The *ulna* is longer stouter, postaxial and more curved than the radius. The proximal end bears a large articular surface for the postaxial condyle of the trochlea of humerus and is further projected into a small blunt *olecranon process.* The distal end articulates with the carpals.

3. Hand or manus

It is differentiated into three parts; *wrist, palm* and *fingers.*

Wrist contains only two irregular bones in adults. A small preaxial *radiale* and a large postaxial *ulnare* articulating with radius of ulna. The palm contains an elongated bone, the *carpo.-metacarpus*. It is formed by the fusion of three *metacarpals* with the distal row of carpals. The *first metacarpal* is the shortest and represented by a small stumpy projection at the proximal end of the composite bone on the preaxial side. The middle or *second metacarpal* is a stout and straight with expanded ends. The third metacarpal is slender and slightly curved joined to the middle carpal at both of its ends.

The fingers contain phalanges which are small bones. There are three clawless fingers. The first digit or pollex consists of single phalanx; the second digit or index has two phalanges and the third digit has only one phalanx. The fourth and fifth fingers are absent.

II. Bones of Hind limb

The hind limbs are devisible into three parts:

(1) *Thigh,* (2) *shank* and (3) *foot.*

1. Thigh

It has only one bone i.e. *femur*. It is differentiated into three parts *head* which is rounded and smooth proximal part, a middle long and curved part is *shaft* and the distal end possesses two prominent *condyles.* Head bears a conical projection on the outer side which is known as *great trochanter*. There is an *intercondylar groove* between the two condyles. *Patella* which is a sesamoid bone is present at knee joint.

2. Shank

It contains two bones, the *tibio-tarsus* and *fibula.* The tibio-tarsus is a stout, very long bone which is formed by the fusion of tibia and proximal row of tarsals. It lies on the inner side. There are two articular facets at its proximal end for the articulation with femur. Its anterior surface bears a

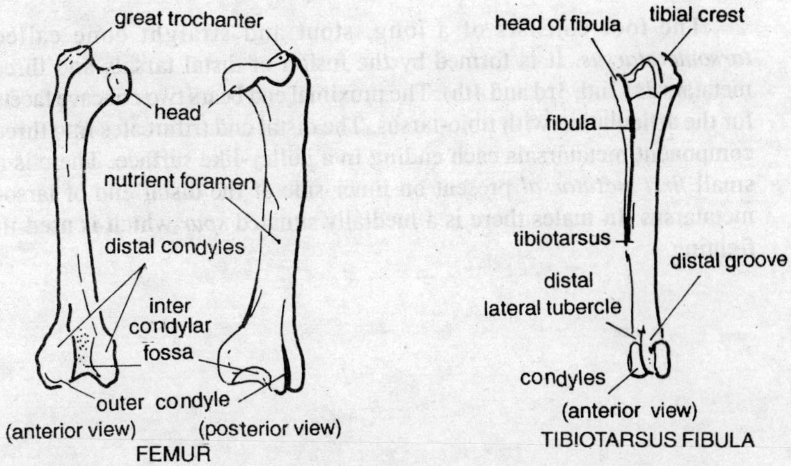

Fig. 4.14. Femur and tibiotarsus fibula.

cnemial crest. The distal end has a pulley like articular surface for the tarsometatarsus. The *fibula* is a small, slender bone which is attached to the postaxial or outer surface of tibio-tarsus. It enlarges proximally and articulates with femur while tapers distially and does not reach the ankle.

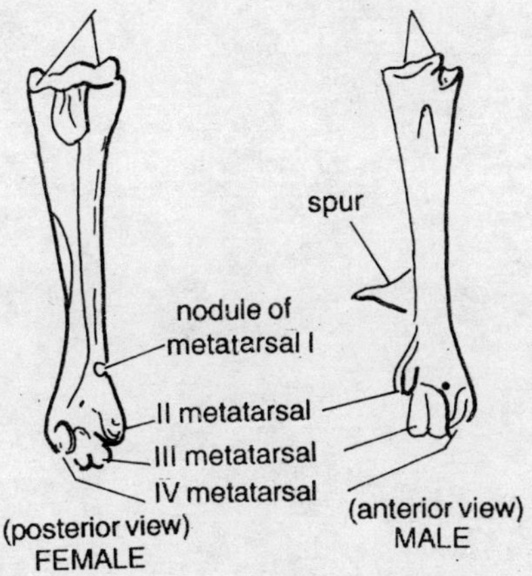

Fig. 4.15. Tarsometatarsus of female and male.

3. Foot

The foot consists of a long, stout and straight bone called *tarsometatarsus*. It is formed by the fusion of distal tarsals and three metatarsals (2nd, 3rd and 4th). The proximal end bears two concave facets for the articulation with tibio-tarsus. The distal end trifurcates into three component metatarsals each ending in a pulley-like surface. There is a small *first metatarsal* present on inner side of the distal end of tarso-metatarsus. In males there is a medially situated *spur* which is used in fighting.

5

Flight Muscles

Birds fly in the air and the forelimbs are modified into the wings. Well developed muscles are needed to move the wings. Hence the musculature of birds is generally highly developed and very efficient. For the movement of wings, the flight muscles are greatly modified. The flight muscles are categorized into three categories:

(1) Pectoral, (2) Accessory and (3) Tensors.

Pectoral Muscles

These are most important muscles consist of *pectoralis major* and *pectoralis minor*.

(i) Pectoralis major

It is the largest and most powerful flight muscle arising ventrally from the keel of sternum or breast bone and the clavicle, one on each side. Anteriorly, the broad and flat tendon of each muscle is inserted on the deltoid ridge of humerus. It is a depressor muscle causing the downstroke of wing when it contracts so that the bird is lifted up in the air. They form $^1/_5$ of the total body weight and are dark red in colour

(ii) Pectoralis minor

These are also called as *supra-coracoideus* or *subclavius* or *deep pectoralis*. They are attached the anterior part of the sternum and lie dorsal to the pectoralis major. Anteriorly it tapers to form a long, thin but strong tendon which passes upwards through the *foramen triosseum* to attach the dorsal surface of the head of humerus. This foramen, which is a gap left between clavicle, coracoid and scapula, acts as a pulley.

It is an elevator muscle, by its contraction the wings are lifted upwards.

Accessory Muscles

Several small muscles also help in elevating and depressing the wings which are called as *accessory muscles. Coraco·brachialis longus* is small triangular muscle lies beneath the pectoral muscles. It is a depressor muscle. It arises from the coacoid and the costal process of the sternum. Its narrow tendon is inserted upon the posterior aspect of the head of humerus. *Coraco-brachialis brevis* is another smaller and narrower muscle lies in front of coraco-brachialis longus. It helps to rotate the wing in the glenoid cavity.

There are certain intrinsic muscles called *biceps* are *triceps* of the upper arm which operate the elbow and perform adjustment during flight. *Extensor carpi radialis* and *extensor carpi ulnaris* also help them in the stretching and folding the wing. Two *brachioradialis* muscles help in medial rotation or radius. Digital muscles also serve to move individual part during flight.

Tensor Muscles

Three tensor mscles namely *tensor longus, tensor brevis,* and *tensor accessorius* keep the pre-patagia fully stretched when the wings are spread out during flight. Another tensor. i.e. *tensor posterius* keep the post-patagia stretched during flight.

PERCHING MECHANISM

Perching is the autonomic clenching of the twig or rod or wire or any such object by the toes of legs in birds. It is brought about by special muscles of the legs and their tendons. The muscles involved are known as *perching muscles* and their arrangement is termed as *perching mechanism*. This is quite autonomic and it enables the bird even to sleep on a twig without falling down.

The perching muscles comprise or two sets of muscles (I) *flexor* and (11) *extensor*.

1. Flexor muscles

These muscles lie on the back of the tibiotarsus and are inserted upon the knee-joint. Their tendons are attached to the phalanges on their lower surface. There are about eight flexor muscles six to anterior toe and remaining two to the hind toe. Followings are the chief muscles.

(a) Ambiens

In majority of birds but not in all, a small slender muscle arises from the ileum and passes through the inner surface of thigh. Its tendon bends under the patella round the knee joint to the opposite side of tibio-tarsus and is connected with the flexor muscles of the 2nd and 3rd toes. It plays a minor role in perching and it can be damaged or cut off without serious loss of perching ability.

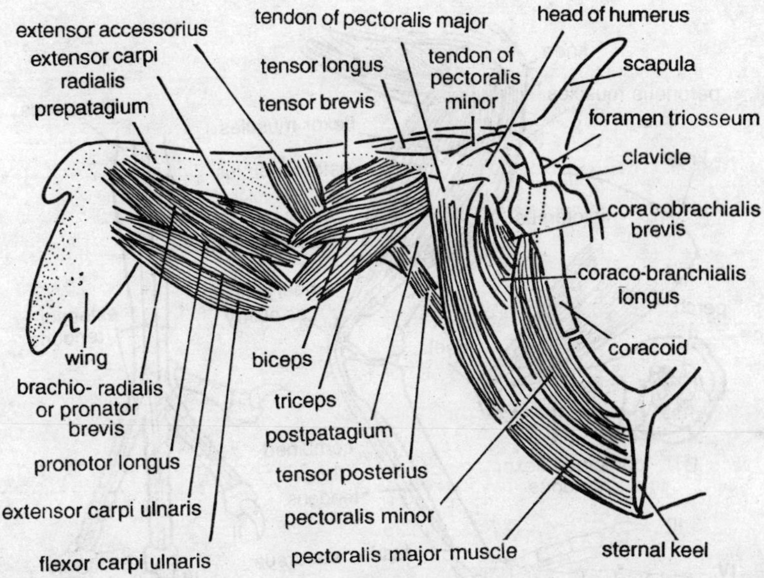

Fig. 5.1. The flight muscles.

(b) Peroneus muscles

These are *longus* and *brevis* attached anteriorly on the crus. Its tendon trifurcates to supply the three anterior toes.

(c) Gastrocnemius

It is the big calf-muscle present on the back of tibiotarsus bone. Its tendon also trifurcates and join those of peroneus muscles to supply the anterior toes.

(d) Flexor perforens

It is also attached to the upper part of crus or tibiotarsus. It gives off a single tendon to the hind toe or *hallux* which is backwardly directed.

2. Extensor muscles

Several extensor muscles are found at the front of tibiotarsus. Their tendons pass down in front of the intertarsal joint and get attached to the upper surface of the phalanges. Their contraction serves to open the toes, when the bird raises its shank while taking off the perch.

MECHANISM OF PERCHING

When the bird rests on a perch, the legs are bent on the knee joint so

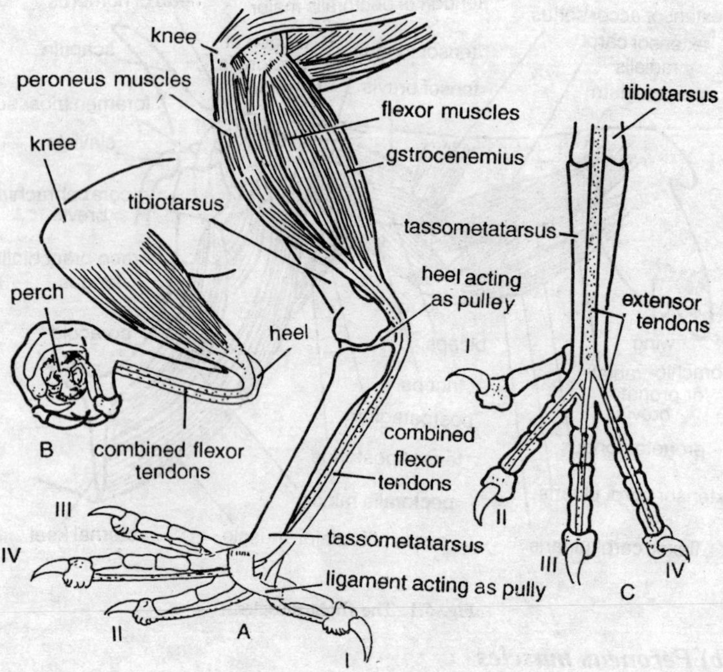

Fig. 5.2. The perching mechanism. A—flexor muscles and tendons in leg. B—Leg bent with toes flexed around a perch. C—Extensor tendons.

that the flexor muscles become stretched. By the pull thus exerted the toes are automatically bent around the perch and grip it firmly. Since all the tendons are interconnected, they act as a single unit, so that pull upon by any one tendon bring about the flexion of all the toes.

The perching is brought about by an inter-locking device. The upper and lower surfaces of tendon sheeths are ridged near the metatarso-phalangeal joints, which is pressed against the perch by mere weight of the body. As soon as the bird settles on a perch, these two opposite sets or ridges interlock so that the toes remain flexed and the bird remains locked to its perch. The toes are open by the contraction of extensor muscles.

6

Digestive System

In birds the digestive system shows great modifications and variations because of large variety of diets adopted by various birds. In pigeons the digestive system can be studied under two headings :

(I) Alimentary canal.

(II) Digestive glands.

I. THE ALIMENTARY CANAL

It is a long and coiled tube of varying diameters. It is divided into following four zones :

(i) *Ingressive zone*. Including mouth and buccal cavity.

(ii) *Progressive zone*. Including pharynx, oesophagus and stomach.

(iii) *Degressive zone*. Includes small intestine.

(iv) *Egressive zone*. Including large intestine.

(i) Ingressive zone

(a) *Mouth.* The opening of mouth is a wide, slit, bounded by upper and lower horny beaks. The beaks do not bear teeth. Mouth leads into a narrow chamber, the *buccal cavity*.

(b) *Buccal cavity.* It is narrow chamber which is somewhat dorsoventrally flattened. It is laterally bounded by the horny beaks, of epidermal origin, which encase the jaws which are without teeth. The roof of the buccal cavity is formed by the skull which bears a pair of elongated palatal folds with free fimbriated margin. It corresponds to the hard palate of mammals, but has a deep palatal fissure in the mid-dorsal line. The palate, therefore, is a split palate. Dorsal to the palatal folds occur the *internal nares.* These are a pair of long, narrow apertures lying side by side in the roof of buccal cavity. The anterior two third of their length is hidden by the palatal folds of mucous membrane. The floor of buccal cavity is occupied by the *tongue*. Tongue is a flat, fleshy organ and triangular with a sharply pointed tip. Posteriorly it is bifid. The tongue

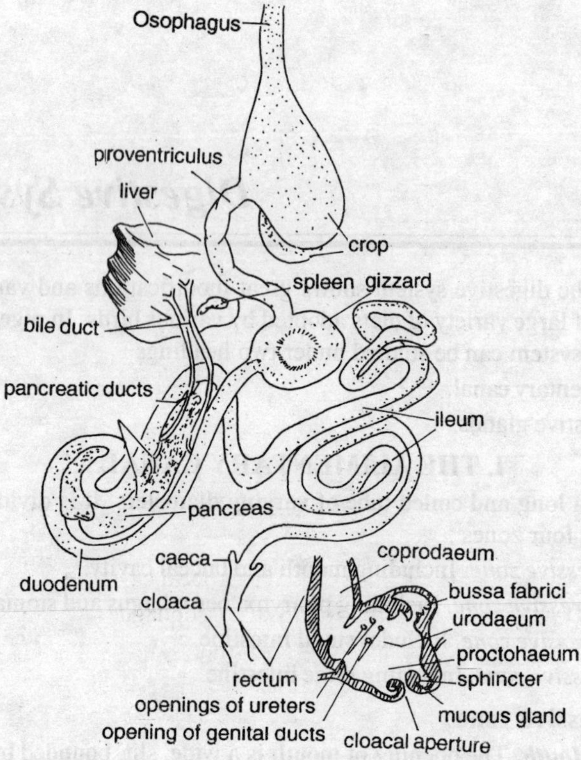

Fig. 6.1. *Columba*. Digestive system; also showing a section of cloaca.

bears few taste buds and glands. The glands include anterior and posterior sublingual glands. In the buccal cavity groups of small glands pour their secretion (which is mucous) for the lubrication of grains which are easily swallowed.

(ii) Progressive zone

(a) ***Pharynx.*** The pharynx extends from the back of the buccal cavity caudally to the point where the trachea begins. In order words it is the portion of the endodermally lined digestive tract which is common to both the digestive and respiratory systems.

On the roof of pharynx behind the internal nares, there is a median aperture known as the *opening of Eustachian* or *pharyngo-tympanic* or *auditory tube*. Each tube extends from this opening to the cavity of middle ear. The floor of pharynx has an oval aperture with tumid lips, the *glottis*, which leads into the trachea. Behind the glottis there is a wide opening, the *gullet*, that leads into oesophagus.

(b) ***Oesophagus.*** The oesophagus is a long tube, which runs back through the neck to the stomach. Peristalsis begins at the oesophagus, which, in most vertebrates, is relatively unspecialized part of alimentary canal between pharynx and stomach. The lining of oesophagus contains mucous gland cells whose secretion lubricates the ingested food particles. The oesophagus has a good deal of elastic tissue, enabling it to stretch enough to accomodate large pieces of food. In most birds the posterior part of oesophagus forms a sac, called, the *crop*, which is particularly large in seed-eating birds like pigeons, finches, buntings etc. In pigeon it is a bilobed and thin walled. The crop primarily is a storage depot, food is eaten rapidly and then passed on gradually for digestion. Here, it may contain over 500 cereal grains for moisten and soften. In pigeon and few other birds, crop glands are formed in breeding season in both the sexes. The crop sloughs off its squameous epithelium to produce pigeon's milk when there is nestling. This milk is regurgitated and fed to young ones. Both males and females produce crop milk, apparently at the mere sight of nestlings. The anterior lobe of pituitary gland secretes a hormone, *prolactin*, which stimulates and controls the formation of crop's milk. The milk contains water, fat, lactose and casein (milk protein). The fat contents are 35% as compared to 3-5% in cow's milk.

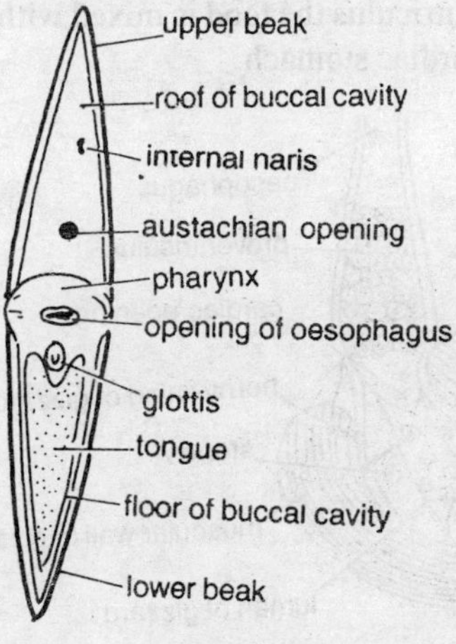

Fig. 6.2. Buccopharyngeal cavity.

The crop contains mucous secreting glands. Beyond the crop,the oesophagus again becomes thick walled, as it enters the stomach.

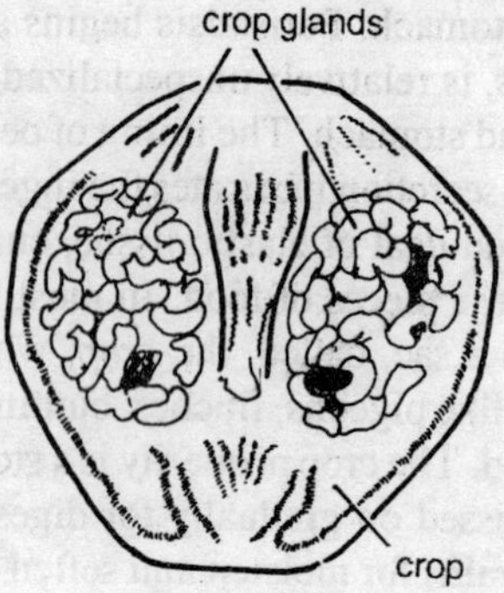

Fig. 6.3. The crop glands.

(c) *Stomach.* The vertebrate stomach is lined with simple columner epithelium which secretes mucous. In most vertebrates it performs three functions, storage, mechanical digestion and chemical digestion.

In birds the true stomach is divided into two parts a glandular *proventriculus* and a *gizzard*. The proventriculus is a small but thick walled sac, externally appearing like a slight dialation of oesophagus. A small, oval and red body, the *spleen,* is attached to proventriculus on right side by peritoneum. In proventriculus the food is mixed with peptic enzymes. Thus, it represents the cardiac stomach.

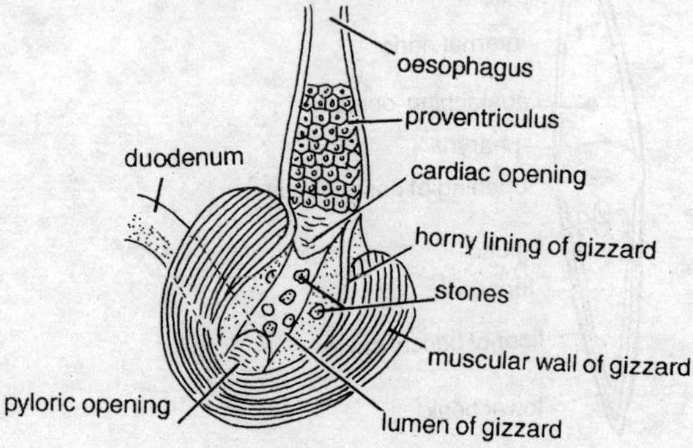

Fig. 6.4. Stomach in sagittal section.

The gizzard has a hard lining of a peculiar material, *koilin*, produced by the secretion of tubular glands, mixed with layers cellular debris. It is green or yellow in colour due to bile. In birds that eat grains or other hard food the koilin may form ridged plates and its action is assisted by small stones in many species. These stones are taken along with the food grains. The gizzard is said to be a substitute for the lack of teeth, appears to be an important adaptation that serves to decrease the weight of head. The gizzard has very great grinding power, its smooth muscles make rhythmic contraction at 1-4 minutes in domestic fowl. *Lazzaro Spallanzani* (1784) reported that a turkey grounds 12 steel needles to piece in 36 hours.

The extent of muscular development depends on both heredity and diet. For example, if herring gulls are fed their usual diet of soft fishes, their gizzard remain thin walled relatively inconspicious. If they are forced to eat grains, however, the gizzard wall soon proliferate and strengthen.

The gizzard represents the pyloric stomach and through pyloric aperture it leads into small intestine. The pyloric aperture is guarded by pyloric sphincter.

(iii) Degressive zone

It includes the *small intestine*. The small intestine is a long and narrow tube measuring 75 cm in length. It can be differentiated into two parts:

(a) *Duodenum* and (b) *Ileum*

The duodenum is anterior part and is'V' shaped. It leaves the gizzard close to the entry of the proventriculus. It receives two bile ducts from liver and three pancreatic ducts from pancreas. It passes into ileum, which is longest part of alimentary canal and highly coiled.

The intestinal epithelium of birds contains columnar and goblet cells similar to those of reptiles, but also has two sharply distinguishing features. In the mucosa are extremely deep grooves, the *crypts of Lieberkuhn*, which carry the epithelium throughout the very thick *lamina properia* and gives the inner surface of intestine a folded appearance. Large branched tubular glands at the base of these crypts secret enzymes which hydrolyse carbohydrates, fats and proteins. The villi are present all over the inner surface of intestine.

At the junction of small and large intestine there are present paired rectal caecae which, in herbivorous birds, may contain symbiotic bacteria and masses of lymphoid tissue. They probably absorb water from food.

(iv) Egressive zone

It includes *large intestine* also known as *rectum*. Rectum is very short being only 4 cm. long. It has the same diameter as the small intestine. The rectum opens behind into the cloaca by way of the anus. The anus is guarded by anal sphincter. In the lining of rectum the number of goblet cells increases while the number of villi decreases.

Cloaca

It is the terminal part of the alimentary canal. It is a large prominent structure and divided into three linear compartments, an anterior *coprodaeum*, a short middle *urodaeum* and a large posterior *proctodaeum*. The coprodacum receives the rectum, urodaeum receives the openings of ureters and genital ducts while the proctodaeum opens to outside by the *vent* or *cloacal aperture*. In young pigeon, there is small blind-sac with much lymphoid tissue, opening into the proctodaeum, it is the *bursa Fabricii*. It is an endodermal derivative. Its function is both to protect locally against infections and to produce lymphocytes for the blood stream, hence it has been called a *cloacal thymus*, as the thymus gland also forms lymphocytes. It atrophies in the adult before sexual maturity.

II. DIGESTIVE GLANDS

The digestive glands include:

(i) Liver
(ii) Pancreas
(iii) Gastric and intestinal glands.

(i) Liver

It is the largest digestive gland, dark-red in colour situated behind the heart. It is bilobed consisting of large right lobe and small left lobe. The surface of liver is hollow to receive the heart, duodenum and gizzard. The gall-bladder is absent in common blue rock pigeon but present in some species of pigeon and birds. Two *bile ducts* arise from the liver, one from left lobe called *left bile duct* and the other from right lobe called *right bile duct*. The left bile duct emerges from the notch between the two lobes of the liver and opens into proximal limb of duodenum. The right bile duct emerges from the dorsal margin of the right lobe of the liver and opens into the distal limb of duodenum. Liver secretes *bile*

(ii) Pancreas

It is a compact, reddish or pinkish gland situated in the loop of the duodenum. There are three pancreatic ducts each opens into the distal limb of duodenum. Pancreas secretes pancreatic juice, containing many enzymes.

(iii) Gastric and intestinal glands

The gastric glands and intestinal glands are microscopic to sub-microscopic in structure. The gastric glands are present in inner lining of proventriculus part of stomach. These glands secrete *gastric juice* containing peptic enzymes. The intestinal glands lie in the mucosa of small intestine. They secrete *succus entericus* or *intestinal juice* containing many enzymes.

Besides the above glands, in pigeon *buccal glands* are also present in buccal cavity, but they do not produce enzymes, they secrete saliva containing mucous for lubrication of food.

Food, Feeding and Digestion

The food requirement of birds is great and digestion is rapid because of high rate of metabolism. An average bird must eats half of its body weight every day. Humming birds must eat far most than this. This animated flying machine requires tremendous energy to operate their powerful breast muscles. Their tremendous metabolic efficiency makes it possible for them to transfer 1/3 of their foodstuff into energy where as mammals can convert only 1/10 of their food into energy.

Pigeons feed almost entirely on the seeds, specially cereals, pulses and sometimes small insects. The seeds are picked up one by one with the help of beaks. The tongue manipulates and lubricates the seeds in mouth cavity with the secretion of buccal glands. The teeth are absent hence the food is swallowed as such into the crop, where it is stored. In the crop the food is softened by the secretion of oesophageal glands and water. The softening in also aided by body warmth. The crop does not secrete any enzyme. The soft food is passed into the proventriculus where digestion is started. The gastric glands of proventriculus produce gastric juice. The partly digested food is passed on to the gizzard where mechanical break down into smaller pieces occur. It takes place by the muscular contraction and stone particles. The ground-up and partly digested food known an *chyme* passes through the pyloric sphincter into the duodenum where it is mixed with bile, pancreatic juice and intestinal juice. Absorption of digested food into blood occurs throughout the small intestine as in mammals. In the rectum the water is absorbed from the faeces, which is passed almost dry.

7

Respiratory System

Because of their high metabolic rate (highest of any vertebrate group) birds require large amount of oxygen and a rapid and efficient exchange of oxygen and carbon dioxide. They have evolved the most extra-ordinary respiratory apparatus. Differing from the usual organization, of vertebrate lungs into airways and gas-exchange surfaces, the bird lung also shows an additional component: its unique system of *air-sacs.* Moreover, true muscular *diaphragm* is absent and the expansion of thoracic wall is limited.

The respiratory system includes the

(A) Respiratory tract

(B) Respiratory organs

(A) RESPIRATORY TRACTS

It includes the following structures

1. External nares

The respiratory tract begins from the *external nares.* These are a pair of slit-like oblique openings lying at the base of the upper beak, just in front of *cere* or *operculum.* The cere is a swollen sensitive pad of skin overlapping the nares. The nares open into *olfactory* or *nasal sacs.*

2. Nasal sacs

They are situated in the olfactory capsules of the skull. They open into the buccopharyngeal cavity by *internal nares.*

3. Internal nares

These are a pair of slit-like openings situated on the roof of buccal cavity, infront of opening of Eustachian tubes.

4. Glottis

It is a median slit-like opening situated on the floor of bucco-pharyngeal cavity just behind the root of the tongue. It leads into the *larynx.*

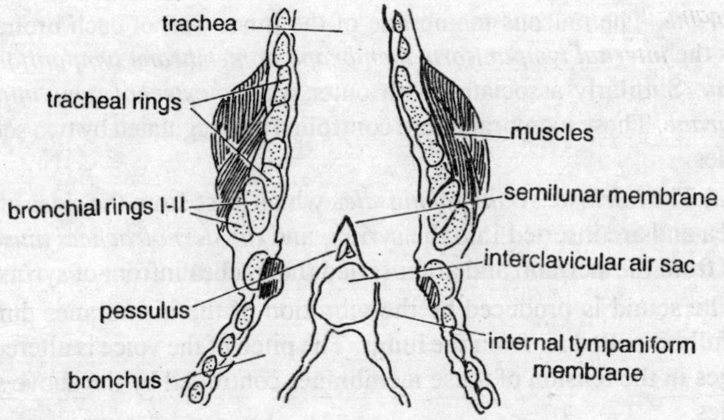

Fig. 7.1. Syrinx in longitudinal section.

5. Larynx

It is a small chamber but it does not function as an organ of sound production, therefore, it is greatly reduced in birds. It is supported by a *cricoid cartilage* (made up of four pieces, the *procricoids*) and by *arytenoid cartilage* (one pair). The *thyroid cartilage* and the *vocal cords* are absent, that is why the larynx does not produce sound in birds. The larynx is followed by wind pipe or the *trachea.*

6. Trachea

It is a long, cylindrical and flexible tube, extends down the neck, ventral to the oesophagus, but is displaced to the left in the middle region by the crop. On entering the thoracic cavity, it expands into a chamber, the *syrinx* and then bifurcates into two bronchi (*primary bronchi*) each entering its respective lung. The wall of the trachea is supported throughout its length by complete *bony rings*. The bronchi are supported by rings, the first one is complete and bony while the rests are incomplete and cartilaginous.

7. Syrinx or Voice box

It is situated where the trachea bifucates into primary bronchi. It acts as a sound box or vocal organ of birds and the characteristic of birds as not seen in other vertebrates. Syrinx consists of an expanded chamber, the *tympanum,* lined by a mucous membrane. Its walls are supported by last three or four rings of trachea and the first half ring of each bronchus. Extended dorsoventrally, at the junction of two bronchi is a bony ridge called *pessulus*. It supports a thin fold of mucous membrane, the *semilunar membrane*. The mucous membrane of the inner wall of each bronchus forms the *internal tympaniform membrane* or *membrane tympaniformis interna*. Similarly associated with outer wall is *external tympaniform membrane*. These membranes are controlled and regulated by two sets of muscles.

(a) The *intrinsic syringeal muscles* which arise from the sides of the trachea and are inserted into the syrinx, and (b) *sternotracheal muscles* arises from the sternum and are inserted the trachea infront of syrinx.

The sound is produced by the vibration of the membranes during forcefully expelled air from the lungs. The pitch of the voice is altered by changes in the tension of these membranes controlled by the above said muscles.

The syrinx is absent in storks and ostriches.

(B) RESPIRATORY ORGANS

The respiratory organs are *lungs* and associated *air-sacs*. The lungs are very small as compared to the size of the body. They are bright red, compact and inelastic structures. They are different from the lungs of other vertebrates as they do not store the air. They do not freely hang in the coelom because they are attached to the thoracic vertebrae and ribs on the dorsal surface hence not covered by peritoneum dorsally. Ventrally they have free surface and remain covered by peritoneum which is modified into thick, strong fibrous sheet called the *pleura* or *pulmonary aponeurosis*. Arise from the junction of the sternal and vertebral ribs, and inserted in the pleural covering, are small fan-like muscles called *costopulmonary* or *intercostal muscles*.

The anterior end of each lung is pierced by a primary bronchus. This bronchus forms a very complex system of respiratory tubules or bronchi inside the lung called *bronchial intercom*. In the lung the bronchus looses its cartilaginous rings and is known as *mesobronchus*. The mesobronchus enlarges into a *vestibule* and extends far back, where it is divided into two branches, one going to abdominal air-sac and other to the thoracic air-sac. From the mesobronchus along its course throughout the lung arise a number of *secondary bronchi* in two sets (a) 4-6 *ventral endobronchi* and (b) 10 *dorsal ectobronchi*. The secondary bronchus rebranched into many fine *tertiary bronchi* or *parabronchi*. Very fine branching and anastomosing tubules radiate from parabronchi called *air-capillaries* or *bronchioles*.

Most part of the lung or lung parenchyma is composed of the central *parabronchus*, surrounded by *air- capillaries* showing hexagonal areas

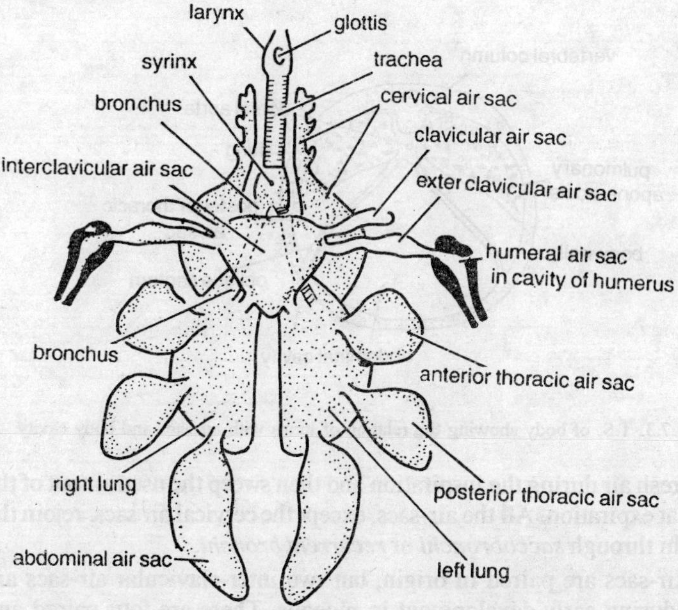

Fig. 7.2. The respiratory system.

or units. Each capillary opens at both the ends into parabronchus. In birds the dead space is absent because the parabronchi and air- capillaries join with their neighbours, none ends blindly. The air capillaries inner surface is highly vascular through which the exchange of gases with blood occurs. The alveoli are absent which are characteristic of mammalian lungs.

About 50% of the lungs volume is occupied by the network of densely packed air and blood capillaries. The size of air-capillaries in birds shows adaptive variation with the metabolism and activity of the species. A pigeon has an air-capillary diameter 3-5 μ and an exchange surface to lung-volume ratio of 302 mm^2/mm^3.

AIR SACS

Associated with lungs are air-sacs. In fact these are the extensions of secondary bronchial tubes except the abdominal air-sacs which arise at the posterior end of mesobronchus. Each air sac is extremely thin-walled, with few nerves and blood vessels and lined with squamous epithelium internally and supporting connective tissue. They are covered over by peritoneum. Their openings in the lungs are called *ostia*. They are filled

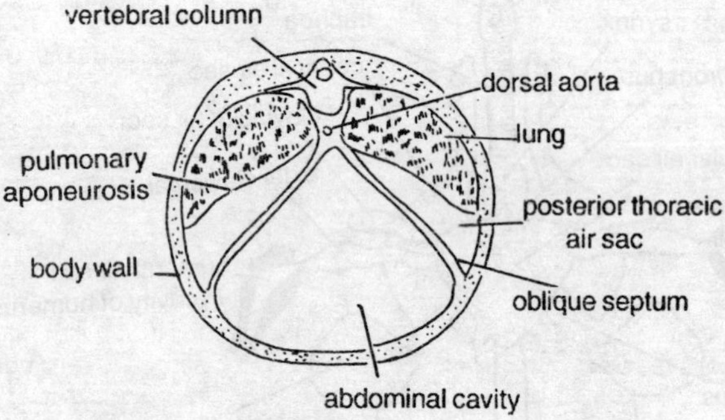

Fig. 7.3. T.S. of body showing the relation of lungs with air sacs and body cavity.

with fresh air during the inspiration and then sweep the used air out of the lungs at expiration. All the air-sacs, except the cervical air sacs, rejoin the bronchi through *saccobronchi* or *recurrent bronchi*.

Air-sacs are paired in origin, but two inter-clavicular air-sacs are fused during early development in pigeons. There are four paired and one unpaired air sacs thus total nine in all. They are named according to their location in the body and are as follows:

1. Interclavicular. It is a median, unpaired air-sac connected with both the lungs. It is triangular in shape and situated between the two limbs of furcula. On either side it gives out an outgrowth the *axillary* or *clavicular air-sac* in the arm pit, of extending further as *extra clavicular air-sac* thence into the humerus bone, where it is called the *humeral air-sac*.

2. Cervical air sacs. These are paired, small air-sacs arising from each lung placed at the root of the neck along with vertebral column. They give off minor saccules in the neck.

3. Anterior thoracic. These are paired air-sacs lie ventral to the lungs in the anterior part of trunk.

4. Posterior thoracic. These are also paired air-sacs found in the posterior part of the trunk just in front of abdominal air-sacs.

5. Abdominal air-sacs. These are paired air-sacs found in the posterior angle of each lung. They lie along the dorsal wall of the abdominal cavity, ventral to the kidneys amongst the coil of small intestine.

The posterior thoracic and abdominal air-sacs are called *inspiratory air-sacs* as the air rushes into these during inspiration. Rest air-sacs are *expiratory air-sacs*.

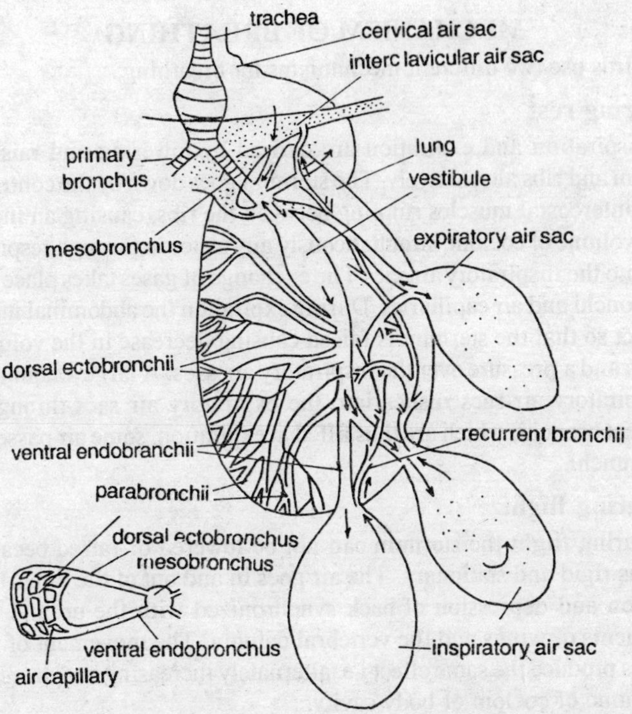

Fig. 6.4. Internal structure of lung showing its relation with air-sacs.

A tough, elastic *oblique septum* covers the ventral walls of the thoracic air-sacs and is continue forward to the pericardium. along the mid-dorsal line, the oblique septum,of two sides unite. Lungs, air-sacs except abdominal sacs lie anterior to the septum. The heart, stomach,liver, intestine and abdominal air-sacs lie posterior to it.

Functions of air sacs

1. They act as balloons providing lightness to the body and reduce the specific gravity of the body as they contain warm air.
2. They act as accessory respiratory organs. They act as reservoir of air which is forced into lungs for ventilation during each expiration. Thus fresh air is available to lungs during inspiration as well as expiration.
3. They may also act as cooling device in the regulation of body temperature. It occurs by internal respiration i.e. water vapours diffuse from blood into the cavities of air-sacs and pass out through the lungs.

4. According to *Muller* (1908) they allow for the movement of heart in a very rigid thorax.

MECHANISM OF BREATHING

Birds use two different mechanisms for breathing:

I. During rest

Inspiration and expiration are carried by lowering and raising of sternum and ribs alternatively. The sternum goes down by the contraction of the intercostal muscles running between the ribs, causing an increase in the volume of coelom, simultaneously air rushes in through respiratory tract into the inspiratory air sacs. The exchange of gases takes place in the parabronchi and air capillaries. During expiration the abdominal muscles contract so that the sternum is raised causing decrease in the volume of coelom and a pressure over the inspiratory air-sacs. A large amount of air of inspiratory air-sacs rushes into the expiratory air sacs through the recurrent bronchi, which are thus filled at expiration, some air passes into endobronchi.

II. During flight

During flight the sternum can not be lowered or raised because it remains rigid and stationary. The air goes in and out of the lungs by the elevation and depression of back synchronized with the up and down movements of wings and the vertebral column. The movement of flight muscles produce the same effect i.e. alternately increasing and decreasing the volume of coelom or body cavity.

8

Circulatory System

Many of features, characteristic of birds, depend on an efficient circulation, allowing a high rate of metabolism, and hence a high and constant temperature. It is significant that the birds and mammals are the only vertebrates that have achieved complete separation of the respiratory and systemic circulations, making possible a high arteriolar pressure, which allows materials to reach the tissue rapidly. The circulatory system is similar as present in other vertebrates and including:

(I) the heart, (II) blood vessels, (III) blood and (IV) the lymphatic vessels.

I. HEART

The heart of birds and mammals show advancement over the ancestral reptilian heart as here it is completely four-chambered with two *auricles* or *atria* and two *ventricles*. Sinus venosus or conus arteriosus is absent in adult bird's heart, the embryonic sinus having merged with right atrium. With the result, the *vena cavae* directly open into the right auricle. The embryonic conus divided into a *pulmonary trunk* joining the right ventricle and a *systemic trunk* joining the left ventricle. Therefore, these arise directly from the ventricles.

External Structure

The heart is a reddish, conical and muscular organ lies in the thoracic cavity, mid-ventrally, partly surrounded by the liver lobes. The size is comparatively larger than that of other vertebrates. Its broad *base* is directed upward and the narrow *apex* directed backward. It is enclosed in a transparent, thin and double membraneous sac, the *pericardium*. The outer membrane is known as *parietal pericardium* and the inner is known as *visceral pericardium*. The two layers enclose a narrow *pericardial cavity*, which is filled with a watery liquid, the *pericardial fluid* or *serous fluid*. The pericardial fluid is secreted by the pericardium. The fluid protects the heart from shocks and injuries, and permits movement during beating.

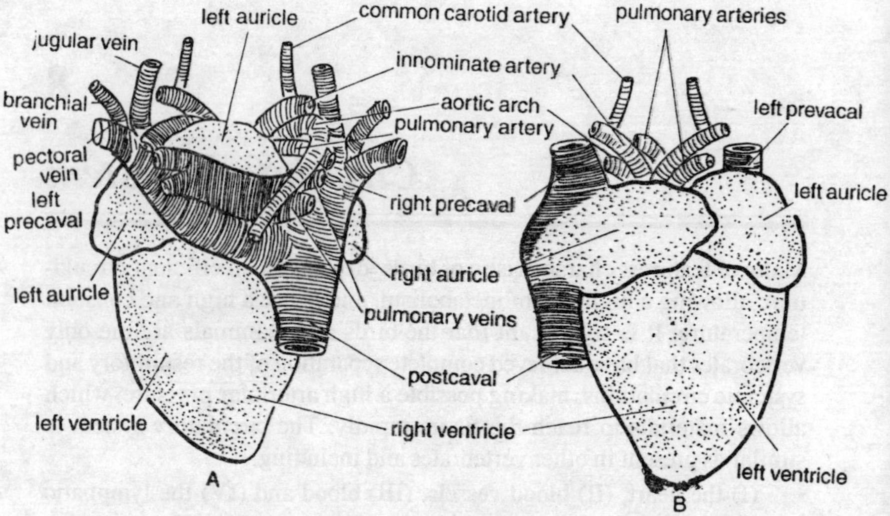

Fig. 8.1. *Columba.* External feature of heart. A—Dorsal view and B—Ventral view.

The two anterior, darker, smaller and thin-walled auricles are separated from the posterior, lighter, bigger and thick walled ventricles by an external hollow transverse groove called *auriculo-ventricular groove* or *coronary sulcus*. A faint *inter-auricular groove* separates two auricles.

Internal structure

Auricles. Internally, the auricles are separated from each other by a thin, muscular, vertical partition, the *interauricular septum*. It bears in its middle a small oval area, the *fossa ovalis,* representing the position of *foramen ovale* in the embryo. The right auricle is comparatively larger than the left and receives the three large cavals, the *right* and *left precavals* and the *post-caval* in its dorsal wall. The opening of post-caval is guarded by a muscular *Eustachian valve,* which projecting into the cavity of the auricle. It checks the back flow of blood from auricle.The other two precavals having no valves.The right auricle opens into the right ventricle by *right auriculo-ventricular aperture*. The aperture is guarded by a pair of strong muscular flaps forming the *right auriculo-ventricular valve*. In function, it represents the *tricuspid valve* of mammalian heart. The chordae tendineae is absent. The left auricle receives four pulmonary veins-from lungs. The left auricle opens into the left ventricle by a circular *left auriculo-ventricular aperture*. It is guarded by two membranous flaps forming the *left auriculo-ventricular valve*. In function in represents the *bicuspid valve* or *mitral valve* of mammalian heart. The posterior free margins of this valve are connected by firm, elastic cords, the *chordae tendineae* to thick muscular processes, the *musculi papillares*, of the ventricular wall.

Ventricles. The ventricles form posterior 2/3 part of the heart. They are paler and stronger than auricles because of thick wall, which are more so in the case of left ventricle. The two ventricles are completely separated from each other by a thick, curved, muscular *intervetricular septum*. The

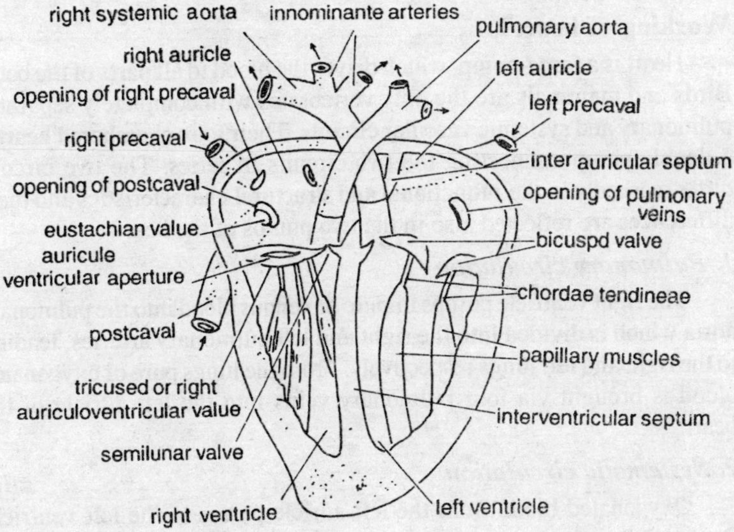

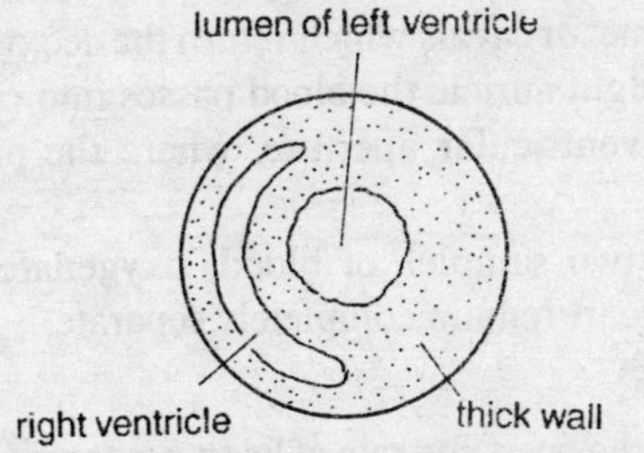

Fig. 8.2. *Columba.* A—Internal structure of heart, B—Ventricle in T.S.

lumen of left ventricle is large and circular whereas the lumen of right ventricle is crescentic. This is because the right ventricle partly encircles the left ventricle. Bars of muscles, called *columnae carneae* or *trabeculae travers* the cavities of the ventricles.

The right ventricle gives off a single trunk, the *pulmonary aorta* which soon divides into two *pulmonary arteries*, each going to a lung. The opening of pulmonary trunk is guarded by three *semilunar valves*. The left ventricle gives rise to the single *aortic arch* or *right systemic arch* which is continued into the dorsal aorta. The opening is guarded by three *semilunar valves*.

Working of heart

Heart is a force-pump, which drives the blood to all parts of the body. Birds and mammals are the only vertebrates with completely separated pulmonary and systemic vascular circuits. Their four-chambered heart is a double pump connecting the two circuits in series. The two circuits differ markedly in their functional and structural characteristics and these differences are reflected also in the two pumps of the heart.

1. Pulmonary circulation

The right ventricle pumps impure or venous blood into the pulmonary aorta which is divided into the right and left pulmonary arteries, leading to the right and left lungs respectively. From the lungs pure or oxygenated blood is brought via four pulmonary veins into the left auricle of the heart.

2. Systematic circulation

Oxygenated blood from the left auricle passes to the left ventricle through left auriculo-ventricular aperture. From the left ventricle sprungs the systemic arch. Through the dorsal aorta, the blood is driven into small arteries, arteries break up into the arterioles and capillaries thus the blood is supplied to all the tissue of the body. The capillaries unite to form venules which unite to form the veins. The veins unite to finally form three venae cavae, or cavals which return the deoxygenated blood to right auricle. From right auricle the blood passes into right ventricle through right auriculo-ventricular aperture, where the pulmonary circulation begins.

Thus, the two samples of blood, oxygenated and deoxygenated coming to the heart remain completely separate.

Heart Beat

The size of the heart and rate of heart-beat varies with size and activity of the birds. Larger birds having in generally smaller and less rapid heart-beats. In a turkey the resting rate of heart beat may be less than 100 beats per minute, in a hen about 300 and a sparrow about 500. Rate increases with exercise and a maximum of 1020 beats per minute have been recorded in humming birds.

BLOOD VESSELS

·Blood travels through the blood vessels throughout the body. These vessels carry pure or impure blood from and to the heart forming (A) The *arterial system* and (B) The *venous system*.

(A) Arterial system

Arterial system carries pure blood to various parts of the body except the pulmonary arteries which carry deoxygenated blood to the lungs. Two large arches arise from the heart: The *pulmonary arch* and *systemic arch*.

1. The Pulmonary arch

It originates from the left anterior corner of the right ventricle, runs forwards between the auricles and divides into the right and left pulmonary arteries. These curve back ward and enter their respective lung.

2. Systemic arch

It originates from the right anterior corner of left ventricle and immediately sends off two small *coronary arteries* to the heart itself. Just in front of heart two short but thick and shout arteries are given off (a) the *left* and *right innominate arteries*. It then curves to the right and crossing the right pulmonary artery, runs upwards, backwards and inwards to reach the mid-dorsal line, from where it continues backwards as the (b) *dorsal aorta*.

(a) ***Innominate arteries.*** The short and stout innominate arteries diverge and each divides into- (i) *Sub-clavian* and (ii) *Common carotid*.

(i) *Sub-clavian*. It extends outwards only for about 3 mm and splits into *brachial* and *pectoral arteries*. The brachial is slender and runs straight outwards to the forelimb. The pectroal artery is thinner than the innominate artery. It breaks up into three branches ramifying in the pectoral muscles. It sends an *internal mammary artery* to the outer wall of thoracic cavity.

(ii) *Common carotid artery*. It runs forward almost parallel to other carotid artery. Each common carotid divides into three arteries: a *vertebral*, an *internal carotid* and a *comes nervi vagi*. The vertebral artery is short and supplies to the neck vertebrae. The internal carotid arteries of two sides run close to each other, then each gives out an external *carotid artery*. The carotids supply the head, its organs, the oesophagus and syrinx.

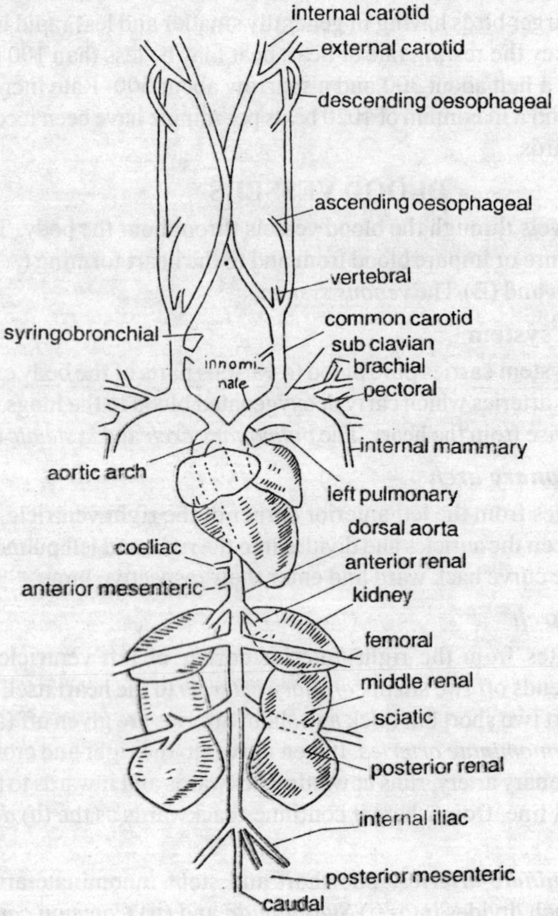

Fig. 8.3. *Columba.* Arterial system.

The comes nervi vagi runs parallel to the vagus nerve and then it joins the external carotid of its side.

The common carotids send *syringo-bronchial arteries* to the syrinx, comes nervi vagi give *ascending oesophageal arteries* to the oesophagus and external carotids provide *descending oesophageal arteries* to the oesophagus.

(b) *Dorsal aorta.* It runs in mid-dorsal line of abdomen and gives off following branches to different body organs.

(i) *Coeliac artery.* A single (unpaired), median, strong artery extends back in the mesentery and gives off branches to proventriculus, gizzard, spleen and liver.

(ii) *Anterior mesenteric artery.* Single, arises about 6 mm. behind the coeliac, runs back in the mesentery and sends branches to small intestine.

(iii) *Anterior renal arteries.* These are a pair of short arteries. They leave the aorta opposite the anterior margins of the kidneys. They enter the first lobe of kidneys. In male each renel artery gives of a branch to the testis, but in female, only the left renal artery gives off a branch to the ovary.

(iv) *Femoral arteries.* These are paired arteries, arise opposite the anterior lobes of kidneys, extend outwards of kidneys and supply the extensor muscles of thigh and pelvic muscles.

(v) *Ischiatic or sciatic arteries.* They are also paired, arising from the aorta next to the femorals. Each passes outwards between the median and posterior lobes of the kidney, to supply the hind-limb.

(vi) *Internal iliac arteries.* These are paired and arise almost from the same place as the posterior mesenteric artery arises. It supplies to the hip region.

(vii) *Posterior mesenteric artery.* It is an unpaired artery and supplies to the rectum and cloaca.

(viii) *Caudal artery.* The dorsal aorta gradually narrows down pasts the kidneys and continues as a small caudal artery into the tail.

(B) Venous system

The Venous system consists of the

1. Caval veins
2. Pulmonary veins
3. Hepatic portal veins
4. Renal portal veins

1. Caval veins

The venous blood returns to the right auricle of the heart through two *precavals* and one *postcaval.* The precavals collect blood from arterior parts of the body while postcaval collects blood from porterior regions of the body.

(a) *Pre-cavals.* The precaval vein or the arterior vena cava of each side is formed by the union of following veins:

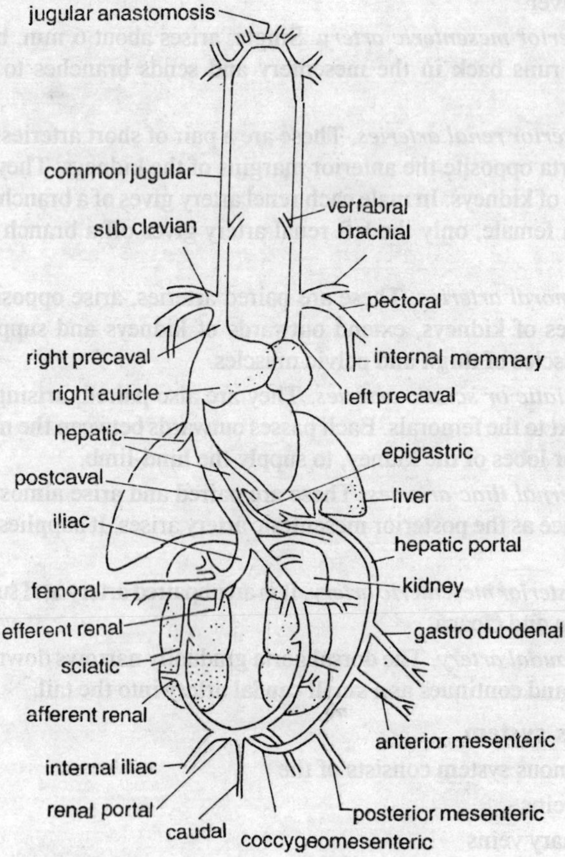

Fig. 8.4. *Columba.* Venous system.

(i) *Common jugular*. Each common jugular vein receives two small veins, the *internal* and *external jugulars,* from the head and neck, and a small vein, the *vertebral vein* from the neck muscles and spinal cord. Two common jugulars are interconnected, at the base of skull, by a transverse loop known as the *jugular anastomosis.* This is an adaptation for shunting the blood from one jugular to the other as the head is turned around and one jugular is blocked.

(ii) *Sub-clavian veins.* Receives a slender vein the *brachial vein* from the wing and a stout *pectoral vein* from the pectoral muscles. A small *internal mammary vein* from the ventral thoracic wall joins the pectoral vein.

(b) ***Post-caval.*** It brings blood from posterior parts of the body including liver. It is formed by the union of two *iliac veins* above the kidneys. The postcaval then enters the right lobe of liver and runs forward. In the liver the post-caval vein receives a few *hepatic veins* and emerges from the anterior side of the liver and then proceeds ahead to open into the posterior side of the right auricle. A small *epigastric vein* arises from the great omentum (a fold of peritoneum lying ventral to the gizzard and intestine) empties into the hepatic vein.

2. Pulmonary veins

Two pulmonary veins carry oxygentated blood from each lung and join with their fellows of the other side to open by a common opening into the left auricle.

3. Hepatic Portal System

As usual the blood from alimentary canal and digestive glands reaches the liver by a *hepatic portal vein* or *median coccygeo-mesenteric vein* which is characteristic of birds. It arises from the point of bifurcation of caudal vein and runs forwards in the mesentery of intestine. It receives three veins, *gastroduodenal, anterior mesenteric* and *posterior mesenteric*. The *gastroduodenal* receives blood from duodenum, gizzard and a part of ileum. *Anterior mesenteric* receives blood from some part of ileum, *posterior mesenteric* from rectum and hinder part of ileum. On reaching the liver the hepatic portal vein divides into two branches, each entering a liver lobe and breaking up into capillaries there.

4. Renal portal system

A renal portal system is practically absent, since very little blood is passed into the kidneys by means of a few *afferent renal veins* and these do not break up into capillaries. *Renal portal veins* are formed due to the bifurcation of *caudal vein*, which is joined by *fermoral* and *sciatic veins* receiving from the hind limbs, and then connects directly with the postcaval. The *caudal veins* receives blood from the tail.

III. Blood

The blood contains R.B.Cs as well as W.B.Cs. The R.B.Cs of birds differ from those of mammals in being oval and nucleated. They are large 10-15μ x 5.5 μ (14μ in pigeon) and their number may be 2.5 - 3.5 million per cubic mm. They carry a large amount of haemoglobin. The R.B.Cs are bigger in flightless birds and smaller in actively flying birds.

Haemopoietic tissue, in adults, is restricted to bone-marrow, it may be formed in spleen and liver.

The number of W.B.Cs is higher than in mammals. They include neutrophils, granulocytes, lymphocytes and monocytes. Thrombocytes also occur but platelets are absent.

Due to higher basal metabolic rate, the temperature of the body is higher, it ranges between 42° - 45° C.

IV. LYMPHATIC SYSTEM

Including few *lymphatic glands* are associated with blood vascular system. Lymplatic tissue is dispersed rather than aggregated into nodes in most birds. There are a pair of *lymph hearts* in the sacral region of embryo and these may persist in adult. The spleen is also intimately connected to lymphatic system. It is a small, oval and reddish gland attached to the right side of proventriculus by a peritoneal fold.

9

Nervous System

The nervous system directs the complex process of the body's internal environment and serves as the link of body with the outside world. All the birds have a nervous system, that together with the endocrine system, acts continuously to preserve personal integrity and a homeostatic steady state regardless of most changes in the external environment.

The nervous system is by far the most complex of all the body systems. Perhaps for this reason it is also the system that is least understood.

The three principal divisions of nervous system are the (I) *Central nervous system,* (II) *Peripheral nervous system* and (III) *Autonomic nervous system.*

The central nervous system consists of brain and spinal cord. The peripheral nervous system consists of cranial nerves and spinal nerves. The autonomic nervous system includes sympathetic and parasympathetic nervous system.

I. CENTRAL NERVOUS SYSTEM

Brain

Birds possess good sense organs and a high sense of intelligence so that the avian brain is roughly ten times the size of that of reptile of the same size and it is comparable with that of mammals.

The brain is short and broad, rounded in form. It is whitish in colour. It is covered over by two protective membranes or meninges, an outer tough *duramater* of connective tissue and inner highly vascular *arachnoid membrane*. Between the two membranes is *cerebrospinal fluid.*

The brain is divided as usual in three parts:

1. Forebrain of *Prosencephalon*
2. Midbrain or *Mesencephalon*
3. Hindbrain or *Rhombencephalon*

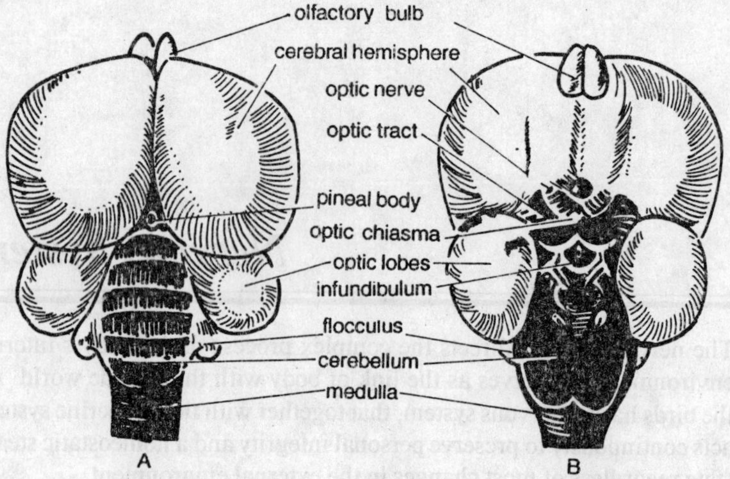

Fig. 9.1. *Columba*. Brain A—Dorsal view. B—Ventral view.

(1) Forebrain or Prosencephalon :

It is divided into following parts :

(a) *Olfactory lobes*. These are conical and very small so that the power of smell is very poor. The cavities or ventricles of olfactory lobes are called *rhinocoel*.

(b) *Cerebral hemispheres*. These are very large and convex. They are separated by a deep sagittal fissure. The cerebral hemispheres are smooth and lacking of folds or fissures characteristic of mammalian brain. The cavities of cerebral hemispheres are called *lateral ventricles* or *paracoels*. The roof or *palliun* is very thin and the *cerebral cortex* is poorly developed. The ventro-lateral walls of cavities are very thick and made up of *corpora striata*. The learning behaviour and ability to adjust in flight are due to enormous development of corpora striata.

An *anterior commissure* interconnects, the corpora striata of two cerebral hemispheres.

(c) *Diencephalon*. It is covered over by cerebral hemispheres on the dorsal side but visible on the ventral side. On the dorsal side just below the median sagittal fissure of cerebral hemisphere lies a single, small, delicate, median *pineal body*. On the ventral side of diencephalon, there is a median projection called *infundibulum* bearing a large *hypophysis* or *pituitary body*. In front of infundibulum is *optic chiasma*, formed by the crossing of two optic nerves coming from optic lobes.

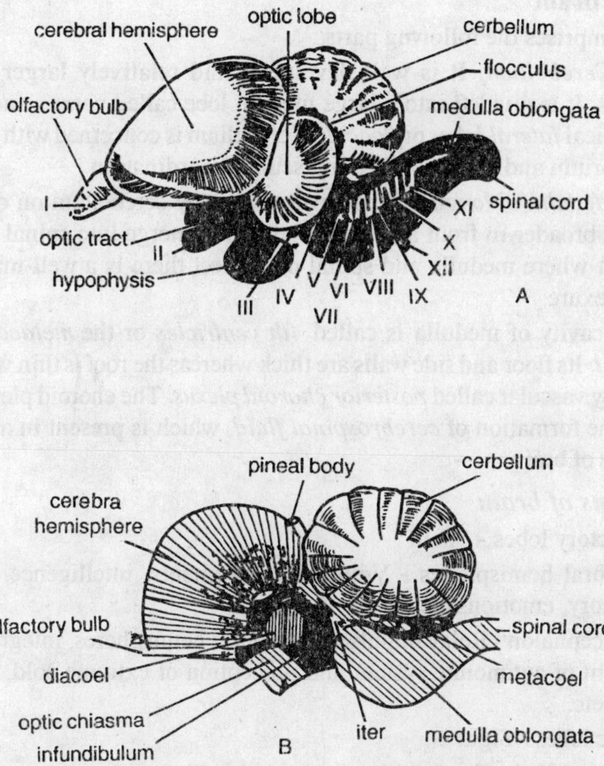

Fig. 9.2. *Columba.* Brain A—Lateral view. B—Sagittal section.

The cavity of diencephalon is called *diacoel* or *third ventricle* connected to paracoel by *foramen of Monro.* Its thin roof is known as *epithalamus* containing a vascular tissue in front of pineal body called *anterior choroid plexus.* The lateral walls are thick and called *optic thalami*, the thick floor is termed *hypothalamus.* A *posterior commissure* separates the diencephalon from mid brain.

2. Mid brain

It is represented by two large, *oval optic lobes*, shifted laterally due to backward extension of cerebral hemispheres. The two optic lobes are connected together by a transverse *optic commissure.*

The cavities of optic lobes are called *optocoels.* The optocoels are connected to diacoel by *iter* or *aqueduct of Sylvius.* Thick longitudinal bands of nerve fibres called *crura cerebri* seen beneath the optic lobes to connect the diencephalon with medulla oblongata.

3. Hind brain

It comprises the followng parts:

(a) *Cerebellum*. It is well developed and relatively larger than mammals. It is divided into a large median lobe called *vermis* and two small conical *lateral lobes* or *flocculi*. Cerebellum is concerned with sense of equilibrium and great power of muscular co-ordination

(b) *Medulla oblongata*. It is covered over by cerebellum on dorsal side. It is broader in front and tapers behind to merge into spinal cord. The point where medulla and spinal cord meet there is a well-marked ventral flexure.

The cavity of medulla is called *4th ventricles* or the *metacoel* or *myelocoel*. Its floor and side walls are thick whereas the roof is thin walled and highly vascular called *posterior choroid plexus*. The choroid plexuses help in the formation of *cerebrospinal fluid*, which is present in all the ventricles of brain.

Functions of brain

1. Olfactory lobes - Smell
2. Cerebral hemispheres - Voluntary movements, intelligence, will, memory, emotions etc.
3. Diencephalon - Relay impulses to cerebral hemispheres, integrating system of autonomic system and preception of extreme cold, heat, pain etc.
4. Optic lobes - Sight
5. Crura cerebri - Conduction of sensory impulses.
6. Cerebellum - Voluntary movement and balance.
7. Medulla - Involuntary movements.

Spinal Cord

The spinal cord is long, whitish, somewhat flattened tube located in the neural canal of the vertebral column. It is covered by meninges as the brain. The *epidural space* and *subdural space* are filled with cerebro-spinal fluid.The epidural space is present outside the durameter and subdural space lies below the durameter. The spinal cord runs the entire length of vertebral column so that there is no filum terminale. In the region of forelimbs and hindlimbs the spinal cord is thickened forming *brachial* and *lumber enlargements*. From these enlargements *brachial* and *sacral plexuses* of nerves are given off. A prominant *ventral fissure* is present throughout the length in mid-ventral line. The dorsal fissure is faint occuring in mid-dorsal line of spinal cord. The spinal cord encloses a narrow *central canal* which opens into myelocoel or fourth ventricle of brain anteriorly but closed posteriorly. Central canal contains cerebrospinal fluid.

Histology

It is made up of two types of nervous tissue, the *grey matter* and *white matter.* Grey matter is inner in position. It is 'H' shaped with prominent *dorsal* and *ventral columns.* The white matter is differentiated into two *dorsal,* two *ventral* and two *lateral funiculi.*

The spinal cord controls reflex activities and serves to conduct impulses to and from the brain.

II. PERIPHERAL NERVOUS SYSTEM

It includes (a) Cranial nerves and (b) Spinal nerve

Cranial nerves

In pigeon there are twelve pairs of cranial nerves which are as follows:

(I) *Olfactory.* They arise from the olfactory lobes and innervates the nasal capsule. They are somatic sensory nerves.

(II) *Optic.* Each optic nerve arises from optic-chaisma situated on ventral side of mid-brain and passes to the orbit of its side. They are somatic sensory nerves.

(III) *Oculomotor.* They are small nerves arise from posterior region of the optic tract in the middle line and supply the orbits. They are somatic motor nerves.

(IV) *Pathetic* or *trochlear.* They arise between the optic lobes and the cerebellum and supply the *superior oblique muscles* of eye. They are mixed nerves.

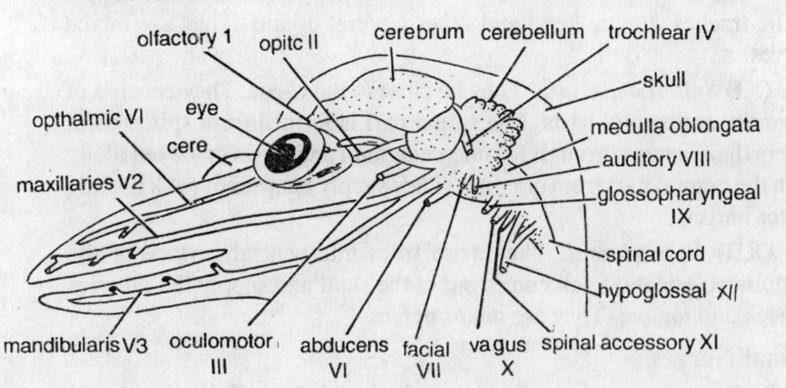

Fig. 9.3. *Columba.* Cranial nerves.

(V) ***Trigeminal.*** Arise from the sides of the medulla behind optic lobes. Each extends upto the *Gasserian ganglion* of its side and divided into there nerves:

V_1 ***Opthalmic.*** It extends along the floor of the skull to reach the orbit and then enters the nasal capsule. It is somatic sensory nerves.

V_2 ***Maxillary.*** After entering the orbit it passes along the floor and goes to the upper jaw. It is somatic sensory nerves

V_3 ***Mandibular.*** It is largest branch of trigerminal. After entering the orbit of its side it divides into two branches one innervate the temporal muscles and other the mandible. It is mixed nerve.

(VI) ***Abducens.*** They are small nerves arise from ventral side of medulla and pass through the skull and supplies to *external rectus muscles* of the eyes. They are somatic motor nerves.

(VII) ***Facial.*** They arise from lateral sides of medulla. Each extends backwards and divides into:

VII_1 ***Palatinus.*** It is visceral sensory formed of fibres from epithelium of the palate.

VII_2 ***Hyomandibular.*** It is mixed nerve supplies to the lower jaw, dermal muscles of face skin and salivary glands.

(VIII) ***Auditory.*** They arise just behind the facial from medulla They supply the ear regions. They are somatic sensory nerves.

(IX) ***Glossopharyngeal.*** They arise from medulla oblongata just behind the VIII cranial nerve. They supply to the mucous membrane of tongue and pharynx. They are mixed nerves.

(X) ***Vagus.*** They arise from medulla. Each runs along the sides of the neck to reach the coelom and divided into many branches which supply to the trachea, lungs, heart and other visceral organs. They are mixed nerves.

(XI) ***Spinal accessory.*** They lie close to the vagus. They consists of a *cranial* and *spinal* parts. The spinal part is formed from spinal cord, enters the cranium through foramen magnum and again come out along with the cranial part from cranium. They supply muscles of neck and are motor nerves.

(XII) ***Hypoglossal.*** They arise from mid-ventral surface of the medulla oblongata. Each comes out of the skull and supply the muscles of neck and tongue. They are motor nerves.

Spinal Nerves

The spinal nerves arise from the spinal cord. In birds the spinal cord is reduced in length due to reduction of the vertebral column. There are

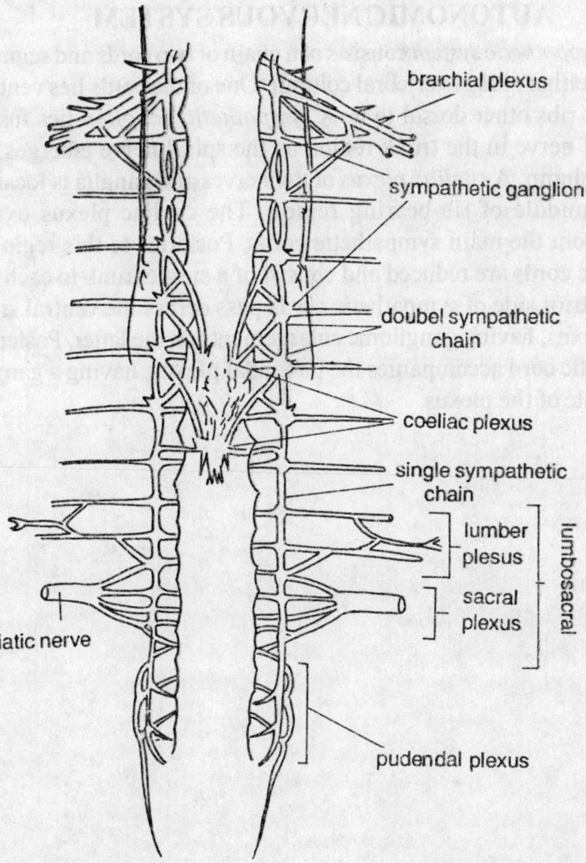

Fig. 9.4. *Columba.* The spinal nerves and sympathetic nervous system.

two enlargements in the spinal cord as in reptiles and it tapers posteriorly. The *cervical spinal nerves* arise at segmental intervals. Their number corresponds with the number of vertebrae in the neck region. A *brachial plexus* is formed by the ventral branch of last cervical and first trunk nerve which is supplied to the wings. Next five pass out between the ribs. After these there is an another plexus, the *lumbosacral plexus*. It is divisible into three parts, the *lumbar, sacral* and *pudendal plexus*. The lumber plexus is formed by three nerves, sending them to thigh. The sacral plexus is formed by five nerves. These five form a large *sciatic nerve* which passes down the leg. Remaining spinal nerves posterior to the sacral plexus form the *pudendal plexus* whose nerves pass obliquely to the cloaca and tail region.

AUTONOMIC NERVOUS SYSTEM

The *sympathetic system* consists of a chain of two cords and segmental ganglia on either side of vertebral column. One of the cords lies ventral to the head of ribs other dorsal to it. A *sympathetic ganglion* lies fused to each spinal nerve in the trunk region as the spiral nerve emerges from vertebral column. A *coeliac plexus* of the nerves and ganglia is located at about the middle of rib-bearing region. The coeliac plexus extends ventrally from the main sympathetic cords. Posterior to this region the sympathetic cords are reduced and consist of a single trunk to each side. On the anterior side of sympathetic cords pass across the ventral side of brachial plexus, having ganglionic enlargements on the latter. Posteriorly a sympathetic cord accompanies the pudendal plexus, having a ganglion in the middle of the plexus.

10

Sense Organs

The sense organs of birds can be classified as follows.

1. Olfactory organs
2. Gustatory organs
3. Tactile organs
4. Auditory organs
5. Photoreceptor organs.

Olfactory organs

The sense of smell is poorly developed in birds that is why the sense organs and the olfactory lobes of brain are reduced. The nostrils are placed at the base of upper part of the beak and are overhung by cere. The nostrils lead into small paired *nasal chambers* or *olfactory sacs*. Both the chambers are separated by the *mesethomoid* and bounded laterally by *ecotoethmoid*. The ectoethmoid is produced inward into three scroll-like processes which increase the surface of olfactory chambers. These scroll-like process are known as *turbinals*. Each chamber has two regions, the anterior is *respiratory part* or *vestibule* and posterior is *olfactory part*. The vestibule containing the anterior turbinal covered by stratified epithelium. The olfactory part containing the middle and posterior turbinals and are lined by single-layered epithelium called *Schneiderian membrane* invested by the fibres of olfactory nerve. Beyond the turbinals the nasal chambers open into phrynax by internal nares.

Gustatory organs

The taste discrimination, which is associated intemately with smell, is also uncertain in birds. The taste buds are few and are usually present in the lining of bucco-pharyngeal cavity. Each taste bud consists of a group of long, narrow *sensory cells* having sensory hair on their free ends and attached to nerve fibres on other end. They are supported by *supporting cells*. Pigeon feeds on grains and seeds make little use of taste because it swallows them.

Tactile organs

The organs of sensitive to touch are poorly developed in birds because of feathery covering of their bodies. The main tactile organs in pigeon is the beak. The *operculum* or *cere* is the most sensitive which is supposed to be an organ of excitement during love making.

Auditory organs

Birds are very much depend on their sense of hearing hence the sense of hearing is acute in most birds. The *pinna* or *external ear* is absent in birds. On either side of head, behind the eye a circular ear opening is present. Each is surrounded by small *auricular feathers* and opens into a short tube called *external auditory meatus*. At the base of auditory meatus lies the *tympanum* or *ear-drum*.

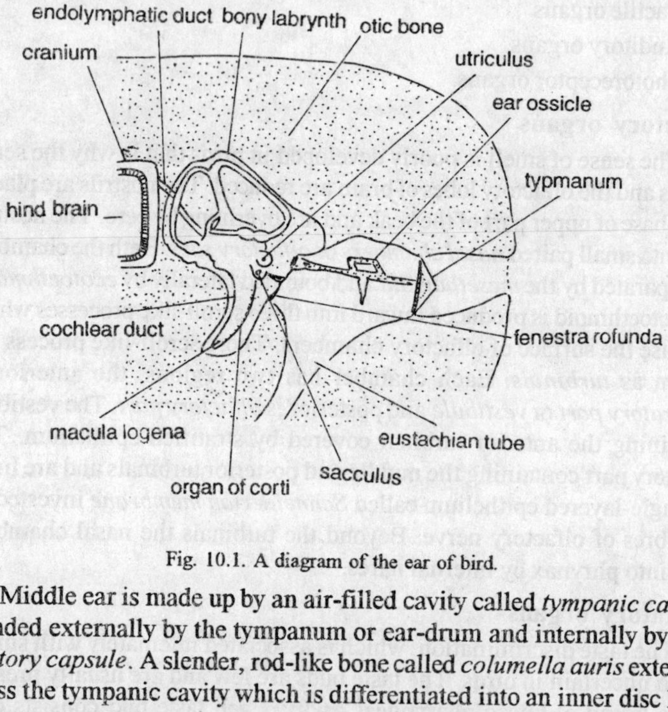

Fig. 10.1. A diagram of the ear of bird.

Middle ear is made up by an air-filled cavity called *tympanic cavity* bounded externally by the tympanum or ear-drum and internally by the *auditory capsule*. A slender, rod-like bone called *columella auris* extends across the tympanic cavity which is differentiated into an inner disc like bony *stapes* and the outer three rayed cartilaginous *extra columella*. The stapes fits into a small membrane bounded aperture, the *fenestra ovalis*, in the outer wall of auditory capsule. The extra columella is attached to inner surface of the tympanum. Below fenestra ovalis is *fenestra rotunda*. It is a membrane-bounded circular aperture. From the tympanic cavity, on each side, an *Eustachian tube* leads to the pharynx and unites with that of the opposite side to open by a common aperture at the back of palate. It serve to equalize the air- pressure on both sides of tympanum.

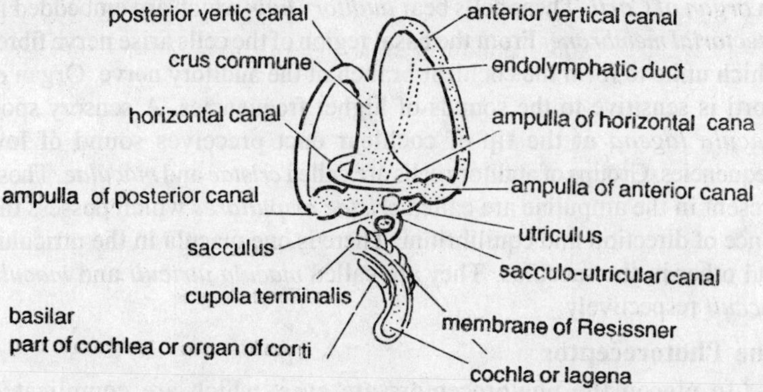

Fig. 10.2. *Columba*. Internal ear (membranous labyrinth).

The internal ear is a soft and delicate structure and is known as *membranous labyrinth*. It is enclosed in a bony labyrinth formed of a dense, ivory-like bone in the side wall of skull. It is surrounded with a fluid, the *perilymph* and itself filled with a dense fluid the *endolymph*. Like other vertebrates it possesses a central chamber the *utriculus*, three *semicircular canals* (one horizontal and two vertical), a small *sacculus* and a short blind tube, the *lagena* or *cochlear duct*. The cochlear duct, in a transverse section, shows three chambers, an upper *scala vestibuli*, a middle *scala media* and a lower *scala typmpani*. Scala media contains endolymph while rest two contain perilymph and continuous with each

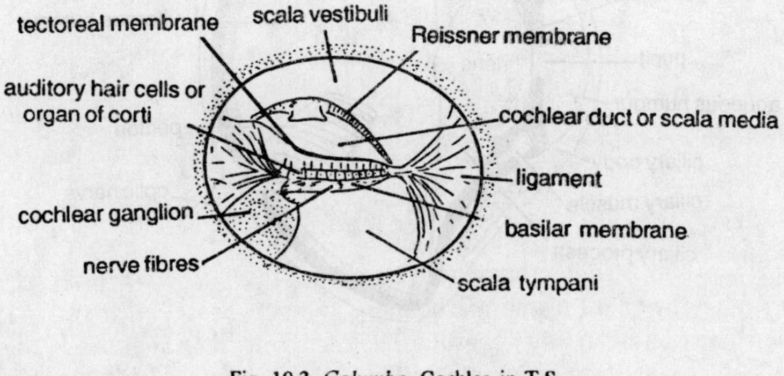

Fig. 10.3. *Columba*. Cochlea in T.S.

other at the tip of cochlca by a passage called *helicotrema*. The floor and roof of scala media are called *basilar membrane* and *Reissner's membrane* respectively. The basilar membrane consists of tall auditory cells forming an *organ of Corti*. These cells bear *auditory hair* which are embedded in a *tectorial membrane*. From the basal region of the cells arise nerve fibres which unite to form the cochlear branch of the auditory nerve. Organ of Corti is sensitive to the sounds of higher frequencies. A sensory spot, *macula lagena* at the tip of cochlear duct preceives sound of low frequencies. Groups of auditory cells are called *cristae* and *maculae*. Those present in the ampullae are called *cristae ampullares* which possess the sence of direction and equilibrium. There is one macula in the utriculus and other in the sacculus. They are called *macula utriculi* and *macula sacculi* respectively.

The Photoreceptor

In pigeon the photoreceptors are *eyes*, which are complicated structures. The eyes are highly developed in birds in adaptation to the aerial mode of life. They are proportionately bigger in size. The structure of the eyes of birds is like that of other vertebrates except a few modifications.

Shape. The eye ball is not a spherical structure as present in man but is partly concave. The concave area is pronounced to varying degrees

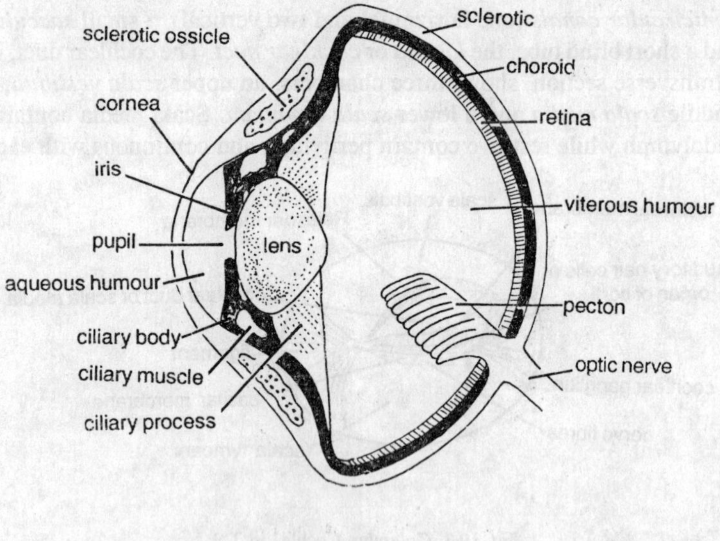

Fig. 10.4. *Columba*. Eye in sagittal section.

and is caused by the presence of a ring of scleral ossicles which surrounds the eye ball in the region of the ciliary apparatus. The portion infront of the ossicles is more or less conical in shape whereas the posterior part is expanded and somewhat flattened.

Eyelids. Two *eyelids* are present, the upper and lower eyelids. The upper is slightly movable but the lower is highly movable. A highly developed *nictitating membrane* or *third eyelid* is present forming a fold at the anterior angle of the eye. It can be swiftly pulled obliquely downwards and backwards over the eye. It is semitransparent and is used to clean and protect the eye ball. In nocturnal birds it reduces the glear of sunlight during day time. In aquatic birds it protects eyes in water and during flight serves as protective covering against the wind. Eyebrows or eyelashes are absent.

Glands. *Harder's gland* or *Harderian gland* is present in the inner angle of the eye. It secerets an oily secretion for lubrication of nictitating membrane. Another glands called tear glands or *lacrymal glands* are also well developed, present below the outer angle of the lower eyelids. They secrete a watery secretion which keeps the eye ball moist and washes away dust particles. This secretion provides nourishment to the non-vascular cornea.

The eye ball is hollow and is made up of three coats (i) outer fibrous tunic (ii) middle uvea and (iii) inner retina.

(i) Fibrous tunic

It is thick and tough and protects the eye ball, maintain the shape and provides surface for the attachment of extrinsic eye-muscles. It is differentiated into two regions anterior small, transparent and exposed portion of the eye beneath the eyelids is *cornea*. The posterior part is large and opaque called *sclerotic* a small portion is visible anteriorly and commonly called white of eye. Sclerotic remains hidden in the orbit. The outer surface of the cornea is covered by a thin transparent epithelium called *conjunctiva* which is fused with outer surface of cornea. The sclerotic is cartilaginous and anteriorly at the junction with cornea, it is strengthened by a ring of 10-12 small overlapping bony *sclerotic plates* or *ossicles*.

(ii) Uvea

It is differentiated into sensory *choroid*, *iris* and *ciliary body*. The choroid coat is a thin, soft, dark brown, vascular membrane, which is adherent to the sclerotic coat. Large and medium sized arteries and veins, as well as numerous capillaries, form a plexus of blood vessels throughout the choroid coat. The pigment serves to keep light from entering the eyeball at random and also prevents internal reflections.

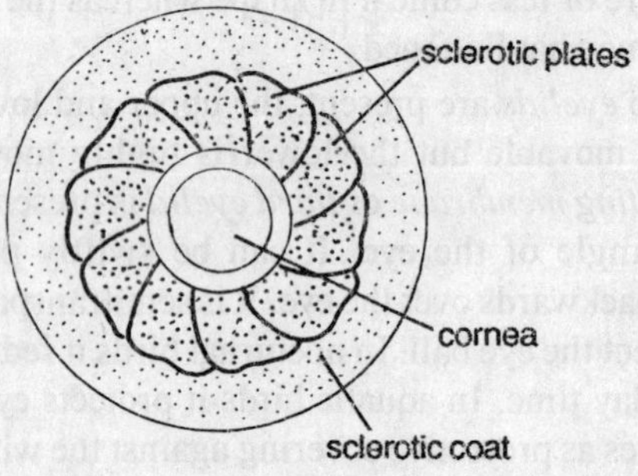

Fig. 10.5. *Columba*. Eye in outer view showing sclerotic plates.

At the point where the choroid portion merges with ciliary body, the uveal membrane becomes thicker, less vascular and less heavily pigmented. Numerous ciliary muscle fibres, which are involuntary, are present in this layer. The inner face of the ciliary body possesses radiating folds, the *ciliary processes*. These are formed partly by the uvea and partly non- nervous border of retina.

Just in front of ciliary body, the uvea abruptly bends into the cavity of eyeball forming a circular partition, the *iris*. Iris is perforated by an aperture, the *pupil*, at its centre. The iris contains two groups of involuntary muscle fibres. The *sphincter* and *dilator muscles* respectively. Contraction of these muscles results in a diminuation in size of the pupil.

(iii) Retina

The pigmented layer of the retina is closely applied to the uveal layer throughout its entire extent. It consists of three regions:

(a) The *optic region* is the thickened portion in contact with choroid coat, sensitive to light. It contains much more *cones* then the *rods* so that there is very clear day light vision.

(b) The *ciliary region* which together with the ciliary portion of the uvea forms the ciliary body.

(c) The *iridial region* contributing to the structure of the iris.

The junction of the optical and ciliary regions of the retina is irregular and called the *ora serrata* or *ora teminalis*.

There are two sensitive spots present on the retina. A true *yellow spot* or *macula lutea* is only present in man and some other primates. In birds *central fovea* or *area centralis* term is used. It lies near the centre of retina as a slight depression and is used for *monocular* or *lateral vision*. Here only cone cells are present and are arranged obliquely rather than perpendicularly. This also may be a factor in bringing about greater visual acuity. The *temporal forvea* lying more towards the outer side of the eye. It is used for *binocular* or *forward vision*. The lower part of retina has yellow and the upper part has red droplets which act as filter, increasing the contrast of blue sky and green fields respectively

Pecten

It is the most interesting feature of bird's eye except kiwi where it is absent. It is a serrated, fan-shaped structure which projects into the cavity of eye from the region of blind spot, where optic nerve enters the eyeball. It is somewhat rectangular highly vascular, pigmented and folded like a fan. The best developed pecten occurs among diurnal birds and the most poorly organized structure occurs in nocturnal birds. The function of pecten has long been the subject of speculation. Even today the actual function of pecten is unknown. Among the most plausible suggestions are:

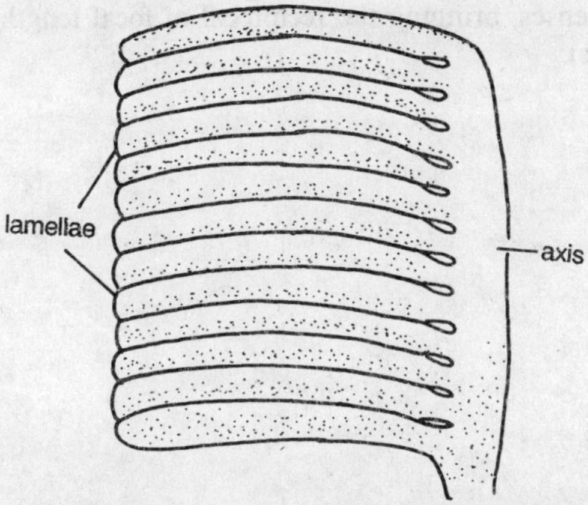

Fig. 10.6. Pecten.

1. that the pecten is an organ of accommodation which by varying the introcular pressure, can cause the lens to change its shape.
2. that it helps in the preception of movements of object because a shadow of pecten falls on retina.
3. that it aids in the nutrition of eye ball.

Lens

A soft, crystalline, colourless, transparent biconvex lens is suspended by suspensory ligaments just behind the iris. It is more convex behind than the front. It is surrounded by a thin membrane, the *lens capsule*. The cavity of eye ball is divided into a small anterior *aqueous chamber* and a large posterior *vitreous chamber* by the lens. The aqueous chamber is filled with a colourless watery fluid, the *aqueous humour*. The posterior chamber contains a thick, colourless, gelatinous substance, the *vitreous humour*. The two humours keep the eye-ball taut and also serve to focus the light rays on the retina.

Accommodation

The eye must adjust itself in order to bring objects at various distance into focus. This is known as *accommodation*. It is brought about by changing the shape of the lens as well as changing the degree of curvature of cornea is also employed. The shape of lens is changed by the striated ciliary muscles. The ciliary muscles are well developed and divided into two sets, the *Crampton's muscle* pulls the cornea reducing its curvature. The *Brucke's muscle* draws the lens forwards increasing the curvature of its anterior surface. Thus bringing about accommodation for near objects. In birds the maximum accommodation is about 20 diopters (a unit adopted for calibration of lenses, bringing the reciprocal of focal length when expressed in metres).

11

Urinogenital System

The excretory and the reproductive organs lie in very close contact structurally, so that they are studied together under the same system i.e. urinogenital system but functionally, they are quite separate.

URINARY OR EXCRETORY SYSTEM

In both sexes, the excretory system is alike and includes one pair of *kidneys* and one pair of *ureters.* A urinary bladder, which is present in other vertebrates is altogether absent due to flight adaptations (to reduce the body weight).

Kidneys

The kidneys are ofcourse metanephric and have a very characteristic shape. Each kidney is large, flattened, three lobed and dark red coloured body filled in the hollow of pelvis. Ventrally they are covered over by peritoneum while dorsally they remain attach to the body wall. A small yellowish elongated and streak like structure lies attached ventrally to the anterior lobe of each kidney, it is known as *adrenal gland.* It is an endocrine gland. Each kidney consists of a large number of convoluted and tightly packed uriniferous tubules. There are 100-500 renal corpuscles in one cubic millimeter of the tissue in cortex of kidney. Each tubule begins as a small glomerulus followed by a long, specialized portion called *loop of Henle.* The reabsorption of water from the glomerular filterate takes here. The kidneys receive a little venous blood from the branch of femoral veins and large amount of arterial blood from the dorsal aorta through renal arteries. The blood leaves the kidneys by renal veins, which unite with femoral and rend portal veins, to form the common iliac veins.

Ureters

The uriniferous tubules of each kidney unite to form a ureter or metanephric duct. The ureters are short, narrow tubes. They arise from the vental surface of the kidneys, between the anterior and middle lobes, run straight backwards over the middle and posterior lobes and open into the urodaeum of cloaca, piercing through the dorsal wall or roof.

Physiology of excretion

The excretory system is highly specialized for water saving. For this purpose the end product of nitrogenous metabolism is relatively insoluble uric acid. Uric acid is synthesized in liver, probably from ammonium lactate. After excretion by the kidneys, the uric acid precipitates in the collecting tubules as whitish granules. The urinary tubules affect a considerable concentration of urine by means of long loop of Henle. The viscous fluid that enters the urodaeum then passes up into coprodaeum and large intestine, where water is further reabsorbed and the mixed faeces and urinary product are then excreted as the characteristic semi-solid white *guano*.

Birds also excrete salt by *nasal glands*, which are well developed in marine forms.

THE REPRODUCTIVE SYSTEM

In pigeons, the two sexes are separate i.e. unisexual, but the sexual dimorphism is absent.

Male reproductive organs

The male reproductive organs include paired *testes* and the *vasa deferentia*. The intromittent or copulatory organ homologous to hemipenis of reptiles or penis of mammals, is absent.

Testes

The testes are large, ovoid bodies varying considerably in size according to the season. The weight of testes is as much as 1000 times greater in the breeding season than it is in the non-breeding period, when they contain only spermatogonia. Each testis remains attached ventrally to the anterior end of kidney by a double fold of peritoneum, the *mesorchium*. The right testis is often slightly smaller than the left one. Each testis consists of coiled, narrow *seminiferous tubules,* lined by the epithelial cells which gives rise to spermatozoa. Groups of *Leydig cells* or *interstitial cells* are present between the tubules. These cells produce hormone. It has been suggested that the abdominal air sacs, which extend into the testicular region help in keeping the two testes cool, which is necessary for the development of spermatozoa, otherwise the high body temperature inhibits the production of sperms, however, there is no definite proof.

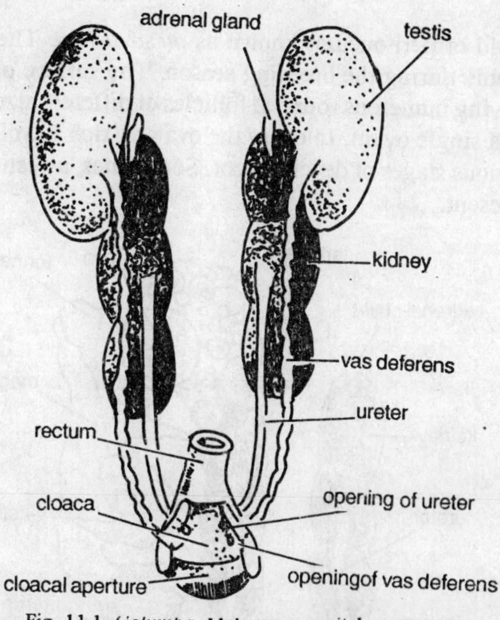

Fig. 11.1. *Columba.* Male urino-genital organs.

Vasa deferentia

From the inner border of each tesatis arises a narrow, opaque, convoluted duct or vas deferens. It runs backwards along the other side and parallel to the ureter of that side, to open dorsally into a small papilla present in the urodaeum of cloaca. The anterior part of each vas deferens is much convoluted to form an *epididymis.* It is connected with the seminiferous tubules of the testis by extreme fine tubules, the *vasa efferentia.* The posterior part of each vas deferens is slightly dilated to form a small *seminal vesicle* or *vesicula seminalis,* for temporary storage of spermatozoa, during the breeding season.

Female reproductive organs

The female reproductive organs developed only on the left side and consists of *ovary* and *oviduct.* In the embryo, however, there are two ovaries and oviducts, but the ovary and oviduct of right side are atrophied in adult, probably as another means of reducing body weight. The left side ovary and oviduct persist and become functional.

(i) Ovary

The left ovary is a large, irregular organ lying ventral to the anterior lobe of the left kidney. Like testis it is also attached to the kidney by a double fold of peritoneum, known as *mesovarium*. The size increases considerably during the breeding season. The surface of ovary is wavy due to having numerous rounded follicles of different sizes. Each follicle contains a single ovum. In ovary the ova are, rich in yolk contents, and are at various stages of development. Sometimes, a vestigial right ovary is also present.

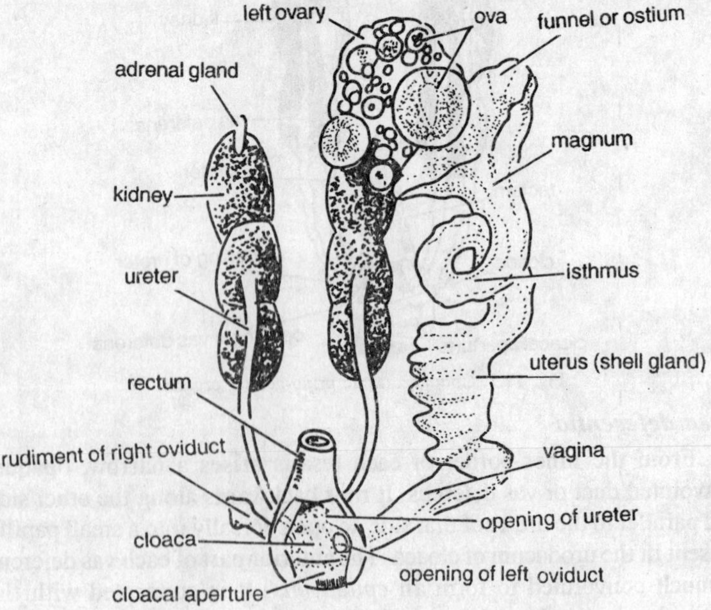

Fig. 11.2. The female urino-genital organs.

(ii) Oviduct

The left oviduct or Mullerian duct is a long, wide, muscular and convoluted tube passing backward to the cloaca. It is attached to the dorsal body wall by a double fold of peritoneum called *mesotubarium*. It is differentiated into following regions.

Ostium or Infundibulum. The upper end of the oviduct forms a wide, membranous coelomic funnel, called ostium, its margin beset with long ciliary processes or fimbrae. It opens by a wide, slit-like aperture into the body cavity, near the ovary. It receives the ripe ova extruded from the follicles. The wall or ostium probably secretes the thin chalaziferous layer of dense albumen around the fertilized eggs.

Glandular part or magnum. It is the next part of oviduct, secretes the albumen or egg white.

Isthmus. Magnum follows a narrow portion called isthmus, which secretes the parchment like shell membranes and some albumen.

Uterus is the 4th part of oviduct, it is thin walled and secretes the hard calcareous shell and albumen.

Vagina. It is the 5th part, and is thick walled and also well muscularised. It probably contains unicellular glands, which secrete the pigments and the ectodermal cuticular layer of shell and also produces mucous for smooth laying of the eggs. The vagina opens into the urodaeum or cloaca.

A vestigial right oviduct is also attached to the right side of the cloaca.

Significance of single ovary

During the embryonic development of most of the flying birds, both the ovaries and oviducts develop, but later on only the left ovary and oviduct are persisted and become functional in adults. This seems to be an advantage to the flying animals, like birds.

The ovary remains small with minute eggs during non-breeding season, but in the breeding season, the ovary and eggs become enormously developed. The eggs not only become numerous but also accumulate the large amount of yolk in them before being laid. Considering the large size of eggs, the birds cannot possibly afford to keep two functional ovaries in their body.

The retention of a single ovary also helps in the safe laying of large eggs with breakable shells. Presence of two ovaries and oviducts might prove disastrous, because if two eggs with breakable shell should approach the cloacal aperture side by side at the same time, the critical stage may be developed.

A single ovary results in loss of weight as well as in decreasing the number of egg production. It is an adaptation for flight.

COPULATION

In brids the copulatory organ like penis or hemipenis is absent, but the fertilization is internal and takes place in the upper part of oviduct (ostium). This follows copulation, which is preceded by courtship. The male treads on the back of female, holding the skin and feathers of the neck of female with his beak. The female lifts up her tail and both press their everted proctodaea closely together. Now the male discharges the sperms into the urodaeum of the female, from urodaeum the sperms are squeezed into the oviduct by contraction of the urodaeum.

The release of eggs from the follicles takes place by the action of luteinizing horomone (LH) from pituitary. In birds there is no corpus luteum. The ovary produces *oestrogens*, *progesterone* and *androgens* probably from the follicular cells in the carly stages of maturation. The lipids and proteins of the yolk are mainly synthesized in the liver and transferred to the egg through the follicular cells. The eggs (usually two) in close succession are directly pass into the oviducal funnel, which is distended and closely applied to the ovary at this time. At this stage each ovum is a mass of yolk with a small germinal vesicle. After fertilization, the ovum rolls down the oviduct in a spiral manner with peristaltic contraction of the wall of the oviduct. During the movement, the albumen and shell are secreted. The albumen is secreted mainly in upper part of oviduct, the lower part producing shell and the lowest mucous, to assist the act of laying. The blue background colour of the egg (oocyanin) is produced shell-formations in the upper part of the oviduct. Spots of red-brown ooporphyrins are added lower down. The pigments are derived from the bile ultimately from haemoglobin.

As much as one-third of the weight of calcium in the whole skeleton, is needed for the shells of the two eggs, laid by a pigeon. The oestrogen they produce, increases the uptake of calcium from the food and stimulates its deposition on the bones. After ovulution the oestrogen level falls, the calcium is mobilized from the bones, probably by parathyroid hormone and its concentration in the blood becomes very high until used by the eggs.

Two eggs are laid at a time in the nest and are incubated at body temperature (41°C) by constant sitting of both the parents i.e. one by one. The incubating period in 2 weeks. The chick breaks the egg shell by *egg-tooth* developed at the tip of the beak. These chicks at this stage are helpless and blind, and are fed by pigeon's milk. This milk is produced by the crop. The *squebs* (unfeathered birds) put on the weight within two days. After about 3 weeks the contours and flight feathers are developed and the nestling are ready to leave the nest. They are taught to fly by the parents.

12

Development of chick

Birds are oviparous and lay their eggs on ground may be on tree in nests. The eggs are of *megalecithal* or *polylecithal type*. The eggs are provided with large amount of yolk. In the avian eggs both yolk and cytoplasm remain separate that is why also called *telolecithal*. The development takes place on ground. Many birds construct nests for their nestlings and some birds like hen, duck etc. lay their eggs directly on ground. The birds, like koel lays eggs in the nest of crow. Most of birds show a high parental care and look after their eggs till they hatch. In the case of hen, the eggs are incubated by the mother. Mother sits over the eggs to develop a suitable temperature for the embryonic development. It takes about 21 days for the complete development of young-one from the egg laid down on ground. To proect the developing embryo, all the basic extra embryonic membranes or foetal membranes like amnion, allantois, yolk sac and chorion are developed.

STRUCTURE OF HEN'S EGG

The eggs of hen are of *telolecithal* type as the yolk particles are present at one end of the egg. As good amount of yolk is present so a term megalecithal or polylecithal can easily be applied. There is a complete separation of cytoplasm from yolk. Cytoplasm, which is confined to a small area the *blastodisc* or *germinal disc* forming a cap at the animal pole. The *egg nucleus* or *germinal vesicle* is present in the middle of blastodisc. The yolk of hen's egg is not of uniform composition, but consists of a central mass of white yolk forming the *latebra* around which are arranged alternating concentric layers of yellow and white yolk. Yellow part of the yolk is thicker than the white part. A pillar of white yolk, the *neck of latebra*, runs from the centre of the yolk to a point just beneath the blastodisc, where it expands into a disc like area, the *nucleus of Pander*. Chemically, yolk is made of high proportion of fat and low proportion of protein, which results in the production of only small amounts of nitrogenous waste.

During ovulation, the eggs are librated in the oocyte stages. Each oocyte is enveloped by a true *vitelline membrane* which is secreted by oocyte itself. This membrane helps to separate the yolk from the albumen. Vitelline membrane is enveloped by secondary membrane, produced by the secretion of follicular cells of ovary. After ovulation when egg passes down the oviduct, teritary membrane is deposited by the wall of oviduct in the form of several layers of albumen. As already stated that proxiaml glandular part of oviduct secretes thin coats of albumen, while very viscous albuminous layers are coated by the middle part of oviduct, and once again thin layers of albumen are added by the lowest part of the oviduct. By this way the albuminous secretion of oviduct constitutes three layers: thin coat-very viscous coat-thin coat. Part of dense albumen forms twisted, cord like structures the *chalazae.* Calcareous shell is secreted by the uterine wall. An *air space* is also developed at blunt end of egg.

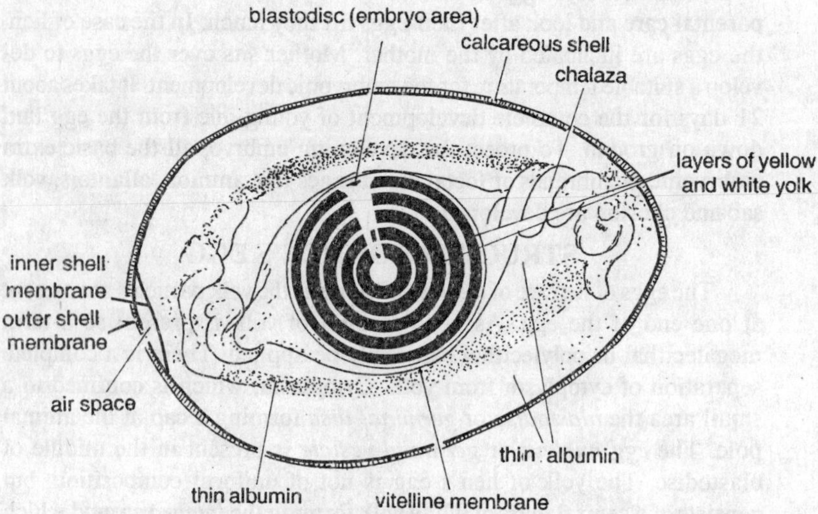

Fig. 12.1. Structure of Hen's egg.

SEGMENTATION OR CLEAVAGE

Segmentation or cleavage starts immediately after the completion of fertilization and continues without interruption as the egg passes down the oviduct. The eggs are laid down about 12 to 16 hrs. after the fertilization. Egg are laid generally at blastula stage and if laying is delayed, development may have gone much further.

The process of segmentation in birds is limited to a small disc due to the presence of mass of the inert yolk material. The disc of protoplasm is present on the yolk material in the animal pole. During cleavage, the entire egg is not divided, so the cleavage is called the *partial* or *meroblastic cleavage*, which is indistinction to the *complete* or *holoblastic clevage*. Holoblastic nature of cleavage can be studied in eggs containing less yolk. A term blastomere can be applied for the cells produced by the process of cleavage.

In the egg of hen, the animal pole is having a disc of active protoplasm which is called the blastodisc. This blastodisc is whitish in colour, circular in outline and about 3 mm. in diameter. This whitish central region of blastodisc is surrounded by somewhat darker marginal area which is called the *periblast.* But actually it is very difficult to make out any line of demarcation between the periblast and white area of blastodisc. Cleavage furrow always starts in the central whitish area of blastodisc while the periblast region remains unaffected till later stages of cleavage are developed.

First cleavage

First cleavage begins soon after the completion of fertilization and continuous without interruption as the egg passes down the oviduct. The nature of the series of divisions is the *discomeroblastic cleavage* characteristic of the eggs of birds. The first cleavage furrow forms in about the middle of the blastodisc, and extends only part ways across it and part way through it. The cleavage furrow always forms at right angle to the long axis of the mitotic spindle. The first cleavage furrow cuts the egg in a plane coinciding with the axis passing through the animal pole and the vegetal pole. The first division or cleavage is completed while passing through the magnum. Two blastomeres are formed by the first cleavage and both the blastomeres remain unseparated from the underlying yolk.

Second cleavage

As the egg enters the isthmus, the second cleavage furrow begins. The second cleavage furrow is developed in the same way but at right angle to the position which was occupied by the first cleavage spindle. This also does not invovle the whole area of the blastoderm, and results in the demarcation of central cells which are surrounded peripherally by a rim of unsegmented cytoplasm (periblast or marginal zone). Later cleavage takes place both in the marginal zone (future area opaca) and in the central area to which are added cells from the dividing marginal zone. By second division four blastomeres are formed.

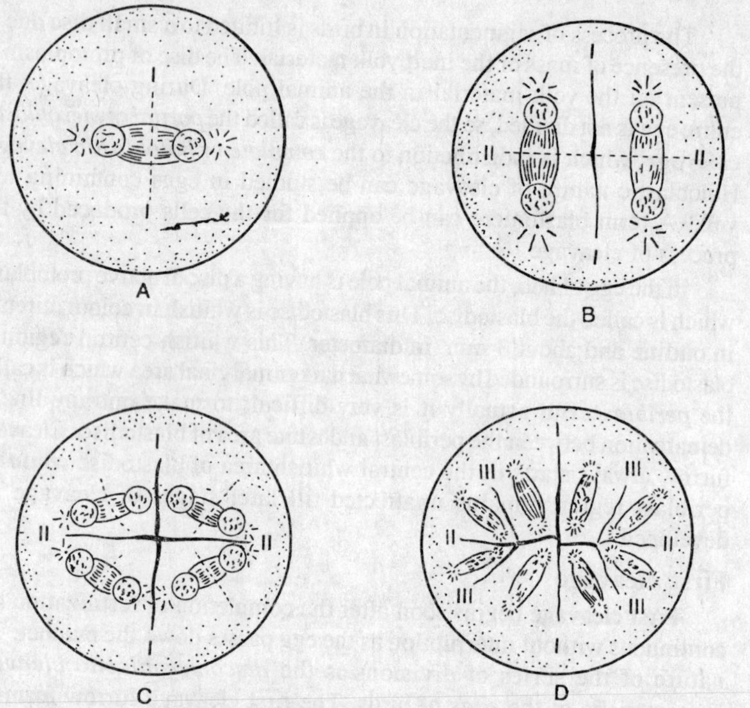

Fig. 12.2. Schematic diagram of cleavage.

Third cleavage

Four blastomeres are developed by second cleavage and during third cleavage in each blastomere cleavage furrow develops. The third division is made up of a series of cleavage furrows which are variable in number and in position. These third cleavage furrows may be parallel with the first, but their direction is quite frequent and irregular.

Third cleavage results in the formation of eight blastomeres. However, the third cleavage appears at first in only two of the blastomeres, so that six cells are formed instead of eight.

Fourth cleavage

Fourth cleavage is also composed of a series of divisions. During this cleavage, all the eight blastomeres produced by third cleavage are cut off from their peripheral portion. Thus the apical region having central cells is separated, their separation is clearly seen in a surface view of the blastoderm but sections show them still unseparated from the underlying yolk. Thus the combined contour of the fourth cleavage furrows form a small irregular circular area.

After the fourth cleavage, the plane of further cleavage becomes irregular. Even though it is possible to trace the cleavage furrows in surface view that divide off additional apical cells. During segmentation radial furrows are also seen that further divide the peripheral cells.

Asscessory cleavage

Generally one ovum is fertilized by one sperm but in birds some times the entry of more than one sperm have been observed. In such caseś when two or more than two sperms enter one ovum, only one of these fuses with the female nucleus and the nuclei of remaining sperms

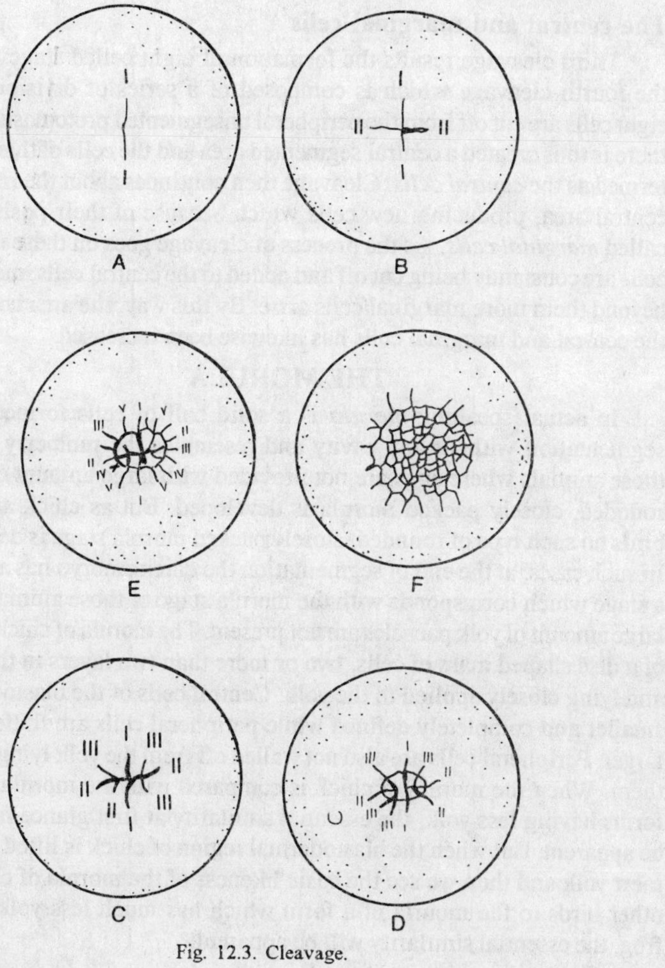

Fig. 12.3. Cleavage.

move to the peripheral part of the blastoderm where they are recognizable for sometime as the so called the *accessory sperm nuclei*. Sometimes these accessory sperm nuclei also undergo division accompanied by the division of the adjacent cytoplasm. The short superficial grooves thus formed are termed as the *accessory cleavage furrows*. Actually no cells are produced by accessory cleavage. After a short duration these sperm nuclei degenerate and accessory cleavage furrows disappear just before the formation of 32 celled stage. These accessory cleavages are of no importance as they do not interfere with the normal development of the embryo.

The central and marginal cells

Third cleavage results the formation of eight celled stage while by the fourth cleavage which is composed of a series of divisions, these eight cells are cut off from the peripheral unsegmented protoplasm. Hence, there is thus created a central segmented area and the cells of this area are termed as the *central cells*. Cleavage then continues about the rim of this central area, producing new cells which because of their positions are called *marginal cells*. As the process of cleavage goes on these marginal cells are constantly being cut off and added to the central cells, meanwhile beyond them more marginal cells arise. By this way, the area covered by the central and marginal cells has likewise been increased.

THE MORULA

In actual speaking *morula* is a solid ball of cells formed during segmentation without any cavity and resembles the mulberry fruit. In those animals where eggs are not provided with large amount of yolk, a rounded, closely packed morula is developed. But as chick and other birds no such type of rounded, closely packed morula stage is developed. In such cases, at the end of segmentation the chick embryo has arrived at a stage which corresponds with the morula stage of those animals where large amount of yolk particles are not present. The morula of chick consists of a disc shaped mass of cells, two or more than two layers in thickness, and lying closely applied to the yolk. Central cells of the blastoderm are smaller and completely defined while peripheral cells are flattened and larger. Peripheral cells are also not walled off from the yolk lying beneath them. When the morula of chick is compared with the morula of other forms having less yolk, the essential similarity at first glance might not be apparent. But when the blastodermal region of chick is lifted from the inert yolk and then we see the basic likeness of the morula of chick and other birds to the mourla of a form which has much less yolk such as frog, the essential similarity will be apparent.

THE BLASTULATION

Morula stage is of short duration and is replaced by the blastula stage where blastocoel or subgerminal cavity develops. As already stated that cleavage or segmentation continuous as long as the egg remains within the body of the hen. After 5th cleavage, however, horizontal cleavage planes under cut the central cells. The cytoplasm which remains beneath these cells then disappears so that the disc of cells, now rightly termed the *blastoderm*, is separated from the underlying yolk by a shallow subgerminal cavity. The nature of the subgerminal cavity also has been a matter of dispute. In the past it has usually been considered to be a blastocoel, homologous with the blastocoel of an amphibian blastula. However, it originates by a disintigration of the cytoplasm and yolk beneath the central cells.

At no time is it actually a cleft between cells as if it is in the amphibian blastula. Soon, however, a true cleft does appear in the midst of the cells of the blastoderm. Some authors consider this to be the true blastocoel.

The zone of junction and the zone of over growth

During this process the margins of the blastoderm continue to grow outwardly across the surface of the yolk. This is done by the thinning of the blastoderm by a process of *proliferation*. By the active cell proliferation and thining of blastoderm the margins of the expanding blastoderm soon develop the yolk so that there is a narrow margin of overgrowth, peripheral to a zone of adhesion between the blastoderm and the underlying yolk.

The egg is laid at about this stage and further development does not take place until incubation has begun. If the egg is not incubated within a few days, regression sets in.

It may now be added that the appearance of this zone of the egg has become a blastula, while the entire cellular and partially cellular area, including the zone of junction and the zone of overgrowth, may henceforth be referred to as the *blastoderm*. The sub-germinal cavity at this stage may be regarded as the equivalent of the blastocoel.

FATE MAP

The plotting of various presumptive or prospective areas is called *fate map*. It is studied by vital stains, carbon particles or radio active thymidine. The presumptive areas of blastula lie within the area pellucida. Beginning from the future posterior end, these areas are: a small disc of *presumptive endoderm*, a broad band of *presumptive lateral plate mesoderm*, two narrow lateral bands of *presumptive somatic mesoderm*, a narrow band of *presumptive notochord*, a large area of *presumptive neural plate* and a very large *presumptive ectoderm*. The latter surrounds the other areas on all sides except the posterior. Outside the embryonic ectoderm is extra embryonic ectoderm lying over the extra embryonic endoderm. The extra-embryonic ectoderm during later developmental stages forms extra embryonic membranes and blood vessels (area vasculosa).

GASTRULATION

The egg is laid at that stage when the differentiation of zone of junction and the zone of overgrowth has taken place. Further development is not possible till the eggs are incubated. So the process of gastrulation can be studied only after the eggs are incubated. As gastrulation begins the blastoderm above the subgerminal cavity becomes thinned by a process of *delamination*. At this stage if an egg is viewed from the surface, the central zone overlying the sub-germinal cavity looks distinct and transluscent and is called the *area pellucida*. In contrast, the surrounding ring like peripheral zone, lying directly on the yolk, looks opaque and white and is called the *area opaca*.

The process of gastrulation involves, at first, the separation of a lower *hypoblast* layer from an upper *epiblast* beginning in the posterior region. The differentiation of epiblast and hypoblast can be studied in following lines:

Epiblast and Hypoblast

As already given that after the differentiation of central and marginal cells, the area increases in size by the multiplication of cells. The area having central and marginal cells is enveloped by a demarcation of slightly darker region which is about 5 mm. in thickness and the same constitutes a ring. This area is called the periblast.

The differentiation of epiblast and hypoblast is taken place from multilayered blastoderm by a process of *delamination*. Smaller cells which are present at the surface and containing less yolk form a thin outer layer the epiblast, and the larger yolky cells move inward from multilayered blastodermal cells by a process of *delamination* and thus they constitute the *hypoblast*. Delamination continuous for a time even after incubation has begun. The space between hypoblast and underlying yolk is the *archenteron*. Some workers believe that hypoblast develops from the individual cells wandering from the surface and detach themsleves within the subgerminal cavity where they eventually become arranged to form a more or less continuous layer.

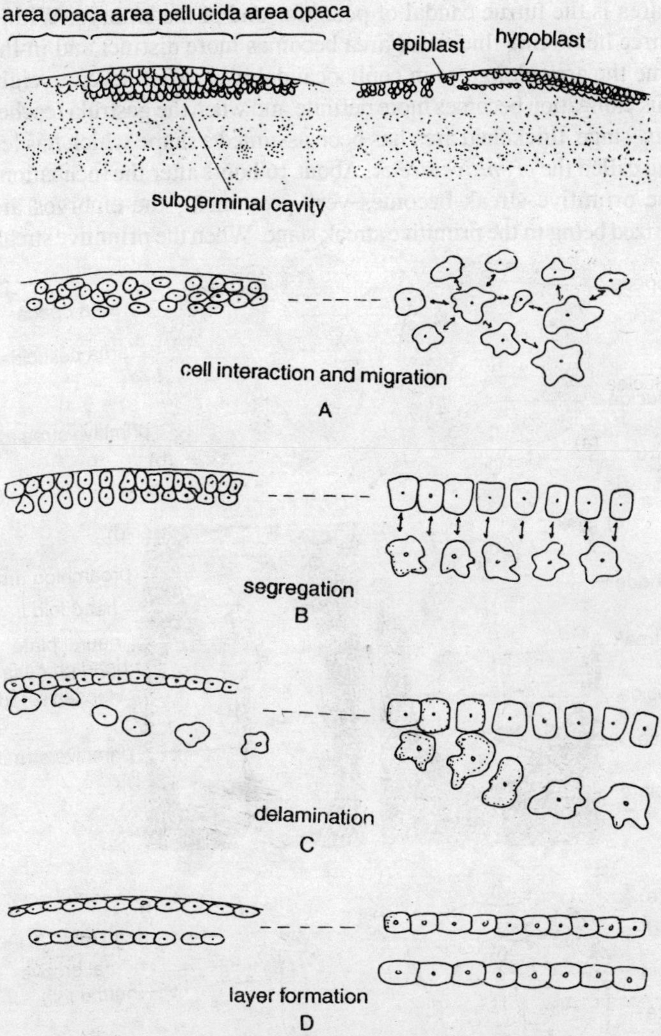

Fig. 12.4. Differentiation of Epiblast and Hypoblast.

The Primitive Streak Formation

After the differentiation of epiblast and hypoblast, the second stage in gastrulation of chick is the development of the primitive streak. In egg that has been incubated for 3 to 4 hrs., the primitive streak begins to form. About three-fourth of the area pellucida becomes more darkly stained definitely thickened as compared with the others. This thickened opaque area is the futrue caudal of posterior end of the embryo. Within two to three hours this thickened area becomes more distinct and in the mean time the area elongates in cephaocaudal way. After seven to eight hours, this elongation becomes more definite and when the gastrula reaches the 12 hrs. stage, thickened area has been assumed a shape which has led to it being called the *primitive streak*. About 16 hours after the incubations when the primitive streak becomes very prominent, the embryos are characterized being in the primitive streak stage. When the primitive streak

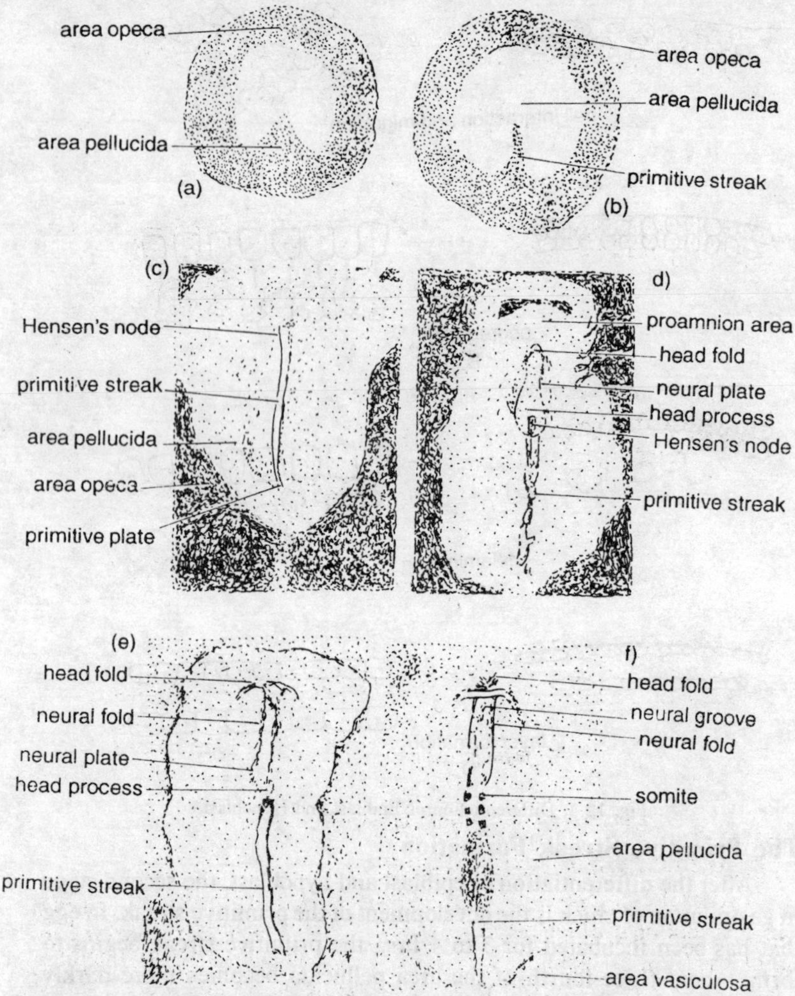

Fig. 12.5. Showing the formation of Primitive streak.

is fully developed, the blastoderm has changed from a circular contour to one that is slightly pear shaped.

A distinct groove develops down as a central furrow, which is called the *primitive groove*. The either sides of the groove are sometimes designated as the *primitive folds* or *primitive ridges*. Towards the cephalic end of the primitive streak, local thickening or swelling is developed by closely packed cells which forms the *primitive knot* or *Henson's node*. The thickness of area pellucida, adjacent to primitive streak, is increased which constitutes the embryonic area. The appearance of this area takes about two hours and can frequently be labelled as *embryonic shield*. The elongation of primitive streak is due to the general stretching of the main embryonic area from where embryo poper is developed while the blastoderm extending beyond its limits is called the *extra embryonic area*. From extra embryonic area, accessory organs are developed which are discarded at hatching.

The origin of the primitive streak

Primitive streak is developed by a process of *convergence* of epiblast cells. During this process the mesodermal cells move towards the middle line of the area pellucida and form a slight ridge, the *primitive streak*, in the posterior region of the blastodisc. After being started, the elongation of the streak occurs. This elongation is done by the stretching of area pellucida in an anterio-posterior direction. It loses its circular shape and acquires a pear shaped out line, the blunt end being anterior. The long axis of the primitive streak definitely establishes the long aixs of the future embryonic body.

Differentiation of ectoderm, endoderm and mesoderm

During gastrulation the cells of blastula arrange themselves in such a precise way that ectoderm, endoderm and mesoderm are developed. Here the formation of these embryonic layers will be discussed individually.

Endoderm formation

In *Amphioxus* and frog, *endoderm* is formed by a process of invagination which is not possible in the egg of hen and other aves due to the presence of enormous yolk. Endoderm is developed in chick by a process of delamination. In blastula the subgerminal cavity is enclosed from the upper side by several cells which are arranged irregularly but during gastrulation from this group of cells some cells proliferate into the subgerminal cavity. After their migration during delamination, they arrange themselves to form a sheet occupying the complete floor of sub-germinal cavity and constitute the embryonic endoderm. During latter stage, the endoderm becomes continuous with the yolk. Actually the cells of hypoblast constitute the endoderm.

Mesoderm formation

After the formation of endoderm the developing embryo contains two definite layers. Out of these, lower being the endoderm and the upper containing the prospective *notochord*, *mesoderm* and *ectoderm*. During the process, the prospective notochord and mesodermal cells move inward and at the end of the process only ectoderm is left over the surface. The

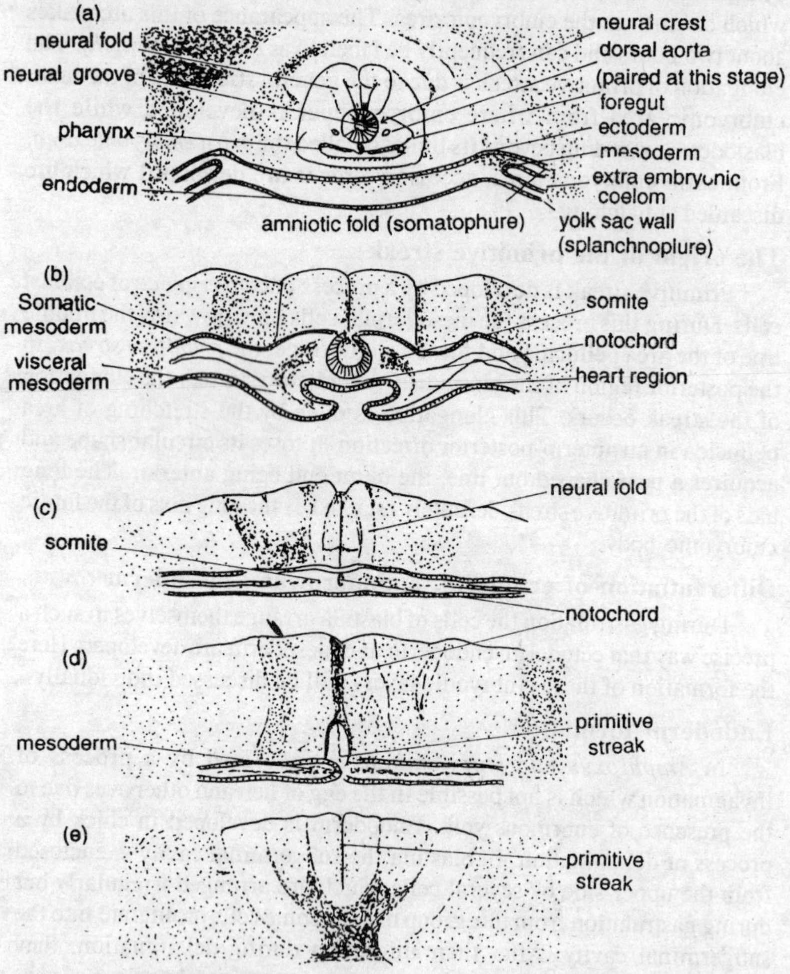

Fig. 12.6. Sections of Chick embryo passing through different regions.

inward movement of cells from upper layers of blastodem, is at first slow and irregular, but soon the movement becomes more organised and much more rapid towards the middle line. This mass movement (*convergence*) of prospective mesodermal and notochordal cells towards the mid-line from a point about a quarter of the way from the anterior end of the area pellucida to a point near its posterior end. The accumulation of cells is seen on the surface of the blastoderm as a dark or opaque streak, the primitive streak. The primitive streak is thus a result of the convergence or crowding towards the mid dorsal line. A further result of the cell movement involved in the formation of the primitive streak, is that the area pelucida loses its circular shape and acquires a pear shaped out line. At the same time primitive groove is developed by two process, they are the *convergence* and *divergence*. During convergence the cells from upper layers are added to primitive streak at its upper borders and by a processes of divergence, the mesodermal cells leave the primitve streak from its lower lateral borders, by this way a groove appears along the length of the streak and this groove is the *primitive groove*. The mesodermal cells leaving the primitive streak during divergence, pass out laterally to form the mesodermal somites on each side between the upper layer of cells and endoderm.

The *somatic mesoderm* begins to differentiate after 21 hours of incubation as dark bands. A cleft appears across each band, a short distance anterior to the primitive node. It marks division between the Ist and IInd somites. Approximately one hour later, a second cleft appears posterior to the first and divides the 2nd somite from the future 3rd somite. The process repeats itself approximately once an hour for 20 hours. After that the somites continue to appear at a somewhat slower rate. The most posterior somites are ofcourse the youngest. The somites are of utmost importance and form the basis for estimating the age of the embryo.

Ectoderm formation

After the separation of endodermal, mesodermal, and notochordal cells from the upper layers of cells, remaining cells constitute the ectoderm. This ectoderm is the remaining portion of epiblast. From ectodermal cells *central nervous system* is formed. Besides central nervous system other organs like intergument, exoskeleton etc., are also developed during *organogenesis*.

COELOM

The coelom in birds is of schizocoelic type. It is true coelom developed from mesoderm. The mesoderm is derived from the source other than archenteon. In the beginning, the mesoderm fans out laterally on either side between endoderm and the upper layer of cells. The solid sheets of this mesoderm extend forward as two horns with a wide gap or true coelom between them. The gap has an upper somatic and lower splanchnic layer of cells. The somatic layer lies next to ectoderm or upper layer of the cells and both are collectively called the *somatopleure*. Similarly the splanchnic mesoderm lies next to endoderm and both the layers collectively called *splanchnopleure*. The solid part of mesoderm gives rise to *dermatome*, *sclerotome* and *myotomes*.

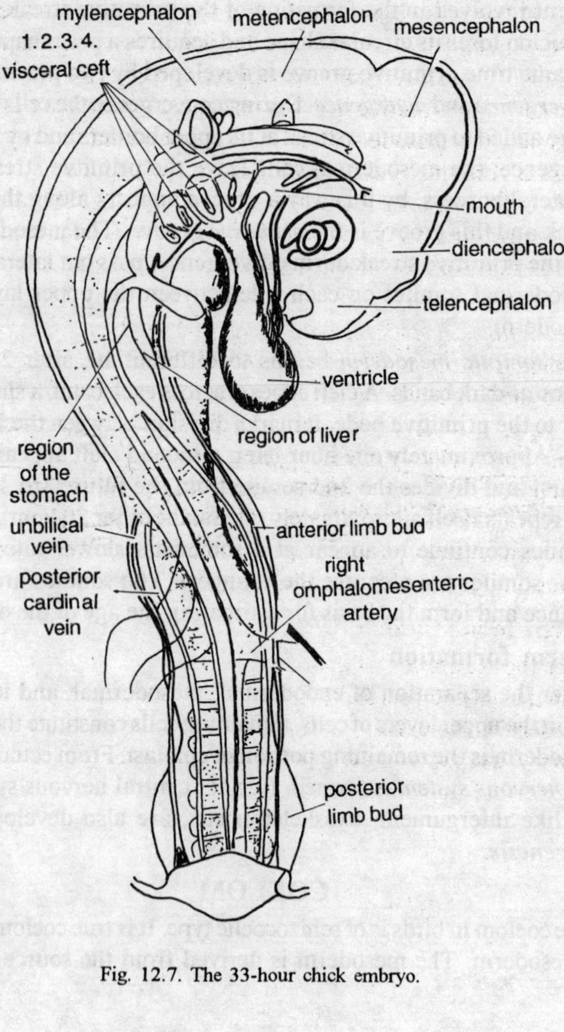

Fig. 12.7. The 33-hour chick embryo.

Development of head process

The embryo proper lies along the longitudinal axis is front of the tail bud or end bud and is surrounded by a large area that does not actually take part in the formation of embryo. The head process consists of a thicker central mass of cells and more diffuse lateral wings. Somatic mesodermal sheets are found on the lateral sides of the notchord. As the development proceeds, the notochord becomes detached from the endoderm (hypoblast) lying immediately behind itself. At this stage the developing embryo shows the continuity of epidermal ectoderm with the neural plate.

The regression of the primitive streak

The formation and development of head process marks the completion of the elongation of the primitive streak. As a result of the streak material moving forward through the centre of the node, the embryonic region in front of the node increases in length, and the primitive streak decreases. The process is termed the *regression of the streak.* In fact, the node migrates backward. The primitive streak becomes gradually shorter and shorter, the small remainder of the node after its regression becomes the tail bud of the embryo. From the *tail bud*, for many hours, the neural tube, notochord and axial mesoderm continue to be proliferated.

FORMATION OF EMBRYONIC AXIAL STRUCTURES

1. Nervous system formation or Neurogenesis

Soon after the formation of head fold, the prospective neural plate begins to give rise the *neural tube,* dorsal to the notochord. The plate can be seen in about 11 hours embryo. The median portion of the neural plate sinks down as the neural groove and its edges rise up as neural folds. These folds grow toward each other over the neural groove and below the epidermis and finally meet at about 5 somite stage in the middle line to form the neural tube. The process of fusion first occurs just in front of the middle of the dorsal surface which then extends forward and backward. For a time, until about 13 somite stage, an opening remains at the anterior end of neural tube known as *anterior neuropore*. At the posterior end, the neural folds remain open for a considerably longer time. Because of its shape it is called *sinus rhomboidalis* or simply the *posterior neuropore*. At about 20 somite stage it is closed. After the formation of the neural tube three primary cerebral vesicles are formed at its anterior end. On the second day, the fore-and mid-brain become differentiated due to appearance of cranial flexure. Hind-brain is differentiated later on by *cervical flexure*.

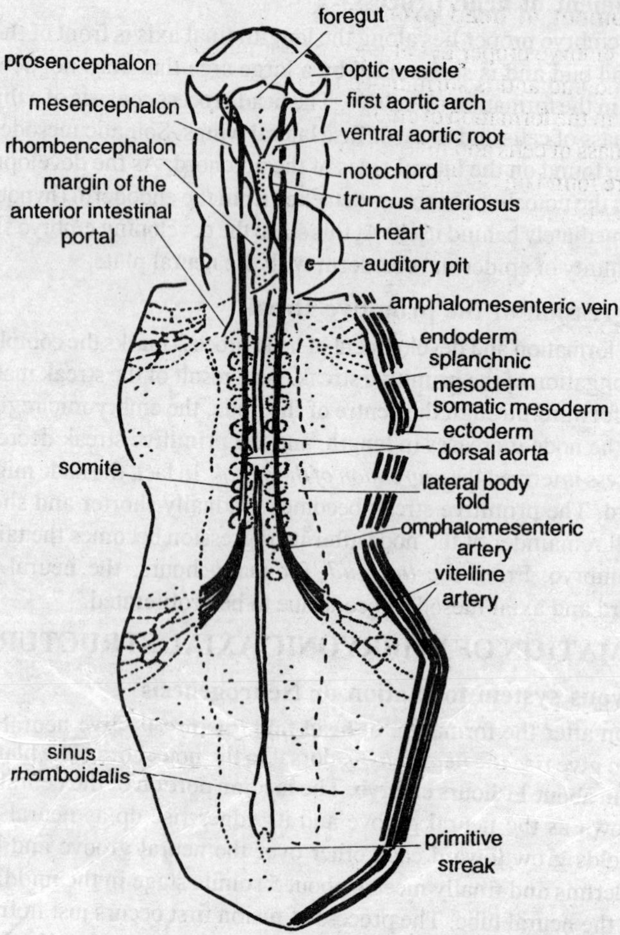

Fig. 12.8. The 72-hour chick embryo.

The primary optic vesicle grows out on either side from embryonic mid-brain at about 12 somite stage or 32 hours of incubation. Each vesicle becomes cup shaped by invagination forming optic cup and its opening represents the *pupil*. Its wall later develops *retina*. Lens develops later on.

Ears develop, when the 15 somite (35 hr. of incubation) have been formed, as a pair of *auditory placodes* behind the optic region. They soon invaginate to form the *auditory pits* which are then converted into *auditory sacs*.

2. Notochord formation or Notogenesis

During the process, the chorda-mesodermal cells migrate forwards forming the primitive node and differentiated into the central tongue of cells known as the *head process*. The axis of head process becomes the notochord. Notochord forming chords cells become separated from mesoderm and are arranged compactly. The developing structure is then surrounded by a sheath of mesenchymal cells derived from sclerotomes.

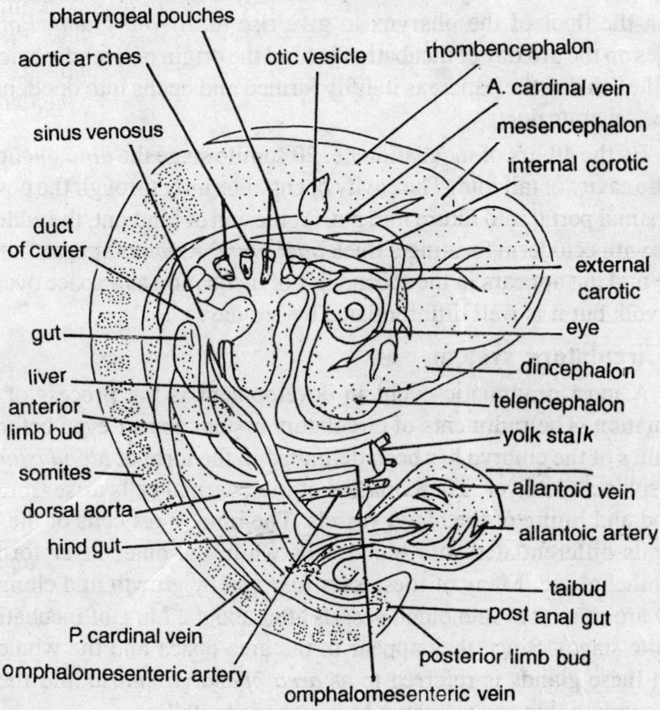

Fig. 12.9. 96-hours chick embryo.

3. Alimentary canal

The alimentary canal first appears in the form of *archenteron*. The endodermal layer is present in the dorsal part of archenteron and the yolk in its ventral part. As the development continued, the foregut appears at the head end of the embryo on the first day of incubation as blind pocket of splanchnoplure (endoderm + splanchnic mesoderm). This at first opens at its sides, and posteriorly on the yolk by an opening called *anterior intestinal portal*. Foregut gradually increase in size. At about 32 hrs. of incubation (12 somite stage) broad part of foregut in head fold gives rise to pharynx. The endoderm binding the anterior end soon fuses with the ectoderm ventrally to form *oral plate*. On this plate the *mouth* appears when the embryo is 60 hrs. old.

About 32 hrs. of incubation the pharyngeal wall pushes out laterally as *pharyngeal pouches*. These pouches finally fuse with ectoderm. At the fusion of first three pouches, perforations occur which give rise to *visceral clefts* but the fourth pouch, however, opens outside. Visceral arches appear during the second day. On the second day, a small diverticulum arises from the floor of the pharynx to give rise to *thyroid gland. Pancreas* arises on the 3rd day of incubation behind the origin of liver diverticulum. By the 5th day the pancreas in fully formed and opens into duodenum by three separate ducts.

By the 40 hrs.of incubation i.e. 20 somite stage the *hind gut* appears as the cavity of tail fold. The cavity opens anteriorly through the posterior intestinal portal into future mid gut. At the end of hind gut, the endoderm fuse with ectoderm forming a thick *anal plate*. An *anus* appears later on. The mid gut appears in the second day of incubation as a space overlying the yolk but it is well differentiated by 3rd day.

4. Circulatory system

A most drammatic event in development is the process of heart formation. The rudiments of circulatory system appear even before any somites of the embryo has been delimited in the form of *blood islands* in the splanchnic layer. These masses of mesodermal cells arise from both blood and lining of the blood vessels. The inner most cells of the blood islands differentiated into blood cells while the outer layer forms an endothelial sac. Many of these sacs fuse and by growth and elongation they are converted into blood vessels after about 27 hrs. of incubation (7 somite stage). Since they appear in the area opaca and the whole area with these glands is referred to as *area opaca vasculosa* and the area surrounding this one is known as *area opaca vitellina.*

The blood vessels form network and soon from the network of vessels arise two *vitelline veins* or *omphelo- mesenteric veins.*

By this time the heart has already appeared anteriorly below the fore-gut as thickened tube. From endoderm, the *endocardium* develops which forms the inner lining of heart. The layers of splanchnic mesoderm which enfold the endocardium becomes muscular layer of heart and called *myocardium* and its outer covering the *epicardium*. This tubular heart which is visible in the 30 hrs. embryo soon elonates and becomes curved into a loop. In the heart indication of *truncus arteriosus, auricle* and *ventricle* are evident. In the beginning the valves are absent. At 33rd hours (13 somite stage) the ventricle is complete, but the auricle and sinus venosus are still represented by paired primordia. By the time 20 somites have been delimited i.e. after 40 hrs. the heart begins to beat rhythmically. By this time the valves have developed.

Heart continues infront as *ventral aorta* from which I and II or sometimes III aortic arches are given out during the second day. These aortic arches pass through the gill-slits and join the dorsal aortae. The two dorsal aortae run porteriorly and for sometimes spread over the yolk but later on two aortae fuse to form a single dorsal aorta.

The tube-like heart develops sinus venosus at its porterior end. Two vitelline veins open in it. The sinus venosus is connected laterally with two *ducti Cuvieri, ductus venosus* posteriorly. Soon a pair of *anterior cardinal* and a pair of posterior *cardinal veins* open into ducti Cuvieri. Lateral vitelline veins also appear. *Sinus terminalis* appear all around the area vasculosa in the form of circular blood vessels.

5. Urinogenital organs

The kidneys first appear as *pronephros*. These develop from the nephrostomes on each side during 2nd day of incubation. By the end of second day the rudiments of mesonephric kidneys appear. No genital system has develop so for as these organs are visible on 6th day through their rudiments which appear on 4th day.

FOETAL MEMBRANES IN CHICK

The blastoderm not gives rise to the embryo proper, but also certain structures that lie outside the embyro. These structures are called *foetal membranes* or *extra-embryonic membranes.* These membranes have been developed in connection with cleidoic development.

There are four types of extra-embryonic membranes viz. *yolk sac, amnion, chorion* and *allantois*. All the membranes are composite structures in the sense that they involve two germ layers. The amnion and chorion are composed of extra embyonic ectoderm and somatic layer of mesoderm collectively called *somatoplure*. The yolk sac and allantois are made up of extra-embryonic endoderm and splanchnic mesoderm collectively called *splanchnoplure*.

1. Amnino and chorion

At about 30 hours of incubation, the extra embryonic somatopleure, immediately in front of the head of embryo rises up as the *head fold* of the amnion, consisting essentially of a doubling of somatopleure upon itself. The outer layer of ectoderm and an inner layer of somatic mesoderm together form the *chorion*. The inner wall of the cephalic or head fold has the layers reversed i.e. its outer layer is of somatic mesoderm and inner of ectoderm, these two together form *amnion*. The *extra embryonic coelom* extends between the two walls of the head fold. On the third day of incubation the *tail fold* of embryo springs up from the caudal end and moves anteriorly towards the head fold, above the embryo. Since the tail fold appears late hence it is smaller then that of head fold. Both the folds meet behind the middle of embryo. The point where both the folds meet is called the *seroamniotic connection*. The chorion and amnion are

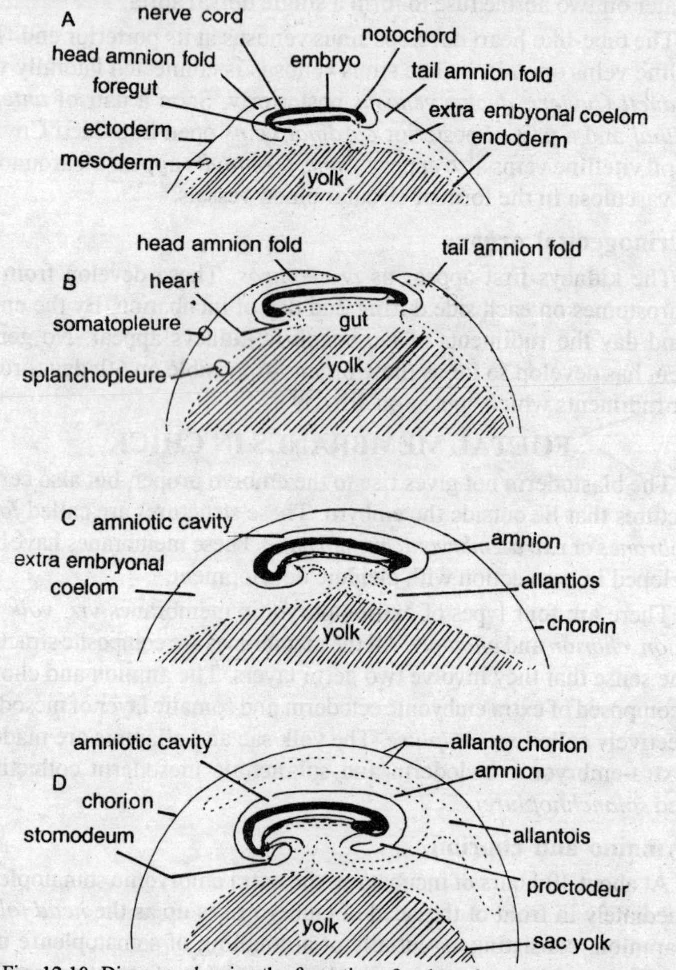

Fig. 12.10. Diagrams showing the formation of embryonic membranes in chick.

separated now. Between the amnion and embryo a cavity is developed known as *amniotic cavity* and it is filled with *amniotic fluid* secreted by amniotic epithelium. The fluid is salty and protects the embryo from external shocks, change in temperature, preventing adhesion of embryo to the shell. By the 5th day in the mesoderm of the amnion unstriped muscle fibres are developed which produce waves of contraction, rocking the embryo gently preventing adhesion and stagnation of blood vessels.

Between the chorion and amnion the *extra embryonic coelom* extends. It is also called *chorionic cavity* and remains filled with a fluid, which provides further protection to the embryo. It also provides space for the extension of allantois. The chorion later becomes pressed against the shell membrane due to excessive extension of allantois. Now the chorion plays an important role in respiration through the porous shell.

2. Yolk sac

In the chick embryo of 16 hours of incubation i.e. the primitive streak stage, the enteron is represented by a flat, circular cavity beneath the premitive streak. Its roof is formed of splanchnopleure and extends peripherally into the extra embryonic area and encloses the yolk mass becoming the yolk sac. The yolk sac remains attached to the midgut of embryo by a narrow *yolk stalk*. The endodermal cells of yolk sac break the yolk for digestion and the digested yolk is provided to the developing embryo. The digestive surface is considerably increased by folds of the wall of yolk sac dipping deep into yolk as the so called *yolk septa*. The digested yolk is reached to embryo by the vitelline veins which open into an unpaired ductus venosus and finally opens into heart. The yolk is completely absorbed before the hatching and the yolk sac is withdrawn into midgut where it degenerates. The yolk sac may thus, be regarded as *extra embryonic gut*.

Besides digestive function, the yolk sac also helps in the formation of blood corpuscles at later stage. Thus it also acts as *haemopoietic organ*.

3. Allantois

After about 60 hours of incubation, a cloacal diverticulum of endoderm grows out from the hind gut of embryo. The splanchnic mesoderm is also pushed out with the diverticulum and forms, with it, the fourth extraembryonic membrane known as *allantois*. Its narrow nack is called the *allantoic stalk*. It grows rapidly, soon extending into the extra-embryonic coelom as a pear - shaped sac and finally comes to lie between amnion and chorion. It is here that the somatic mesoderm of chorion and splanchnic mesoderm of allantois fuse to form a composite membrane, the *allantochorion*. The allantochorion becomes pressed against the shell membranes and gets highly vascularized.

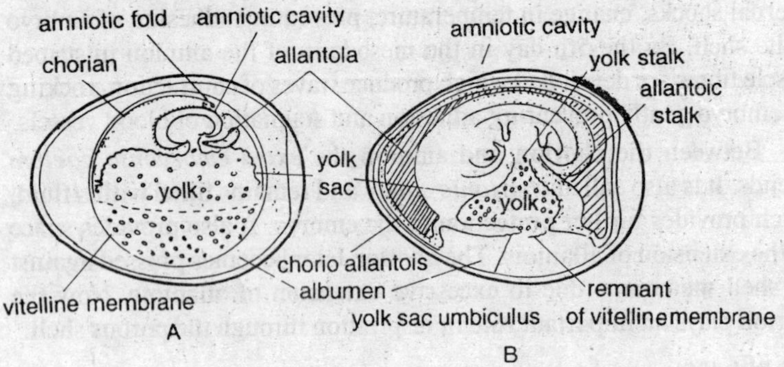

Fig. 12.11. The later development of foetal membranes in chick.

Its capillary network receives blood by a pair of allantoic arteries branching off from the sciatic arteries of the embryo and returns blood by a pair of allantoic veins. The allantochorion later becomes reflected around the albumen so as to enclose it in an *albumen sac*.

The allantoic stalk becomes narrow as development proceeds and is later combined with the yolk stalk in a single outer covering, to form the *umbilical cord.*

At hatching the umbilical cord breaks and the allantios in left behind. The place of attachment of cord to body heals up and a permanent scar is left behind known as *umbilicus.*

The allantois is a reservoir of excretory wastes such as uric acid. The allanto-chorion acts like the lung of embryo as it helps in respiration and moreover the albumen sac aids in absorption of albumen and water.

HATCHING

On 20th day of incubation, the chick ruptures the inner shell membrane adjacent to the air space by means of beak to take the first breath in the air present in the air sac and after this the calcareous shell is broken by sharp horny process, the egg tooth or *caruncle* developed on the upper beak for this purpose. After the break down of shell a little feathery chick emerges out. The newly hatched chick is able to walk and feed for itself. This condition is called *nudifiguous*.

13

Birds are Glorified Reptiles

Nearly a century ago, *T.N. Huxley* stated that birds are *glorified reptiles*, thereby meaning that birds have evolved from some reptilian ancestor. Birds with their power of flight and feathered wings appear to be quite distinct from other groups of vertebrates, yet they have striking similarities in their structure and development with reptiles. On this basis *Huxley* grouped reptiles and birds in one Class-Sauropsida.

The steps and the impulses under which evolution of warm blood birds took place from a cold blooded reptilian stock, are not properly understood at present. The three fold evidences from comparative anatomy, embryology and palaeontology clearly indicate and support the view that birds evolved from thecodont reptiles in early *Mesozoic Era*.

I. ANATOMICAL EVIDENCES

Common features in reptiles and birds are as follows:

1. Exoskeleton

In most of the birds the feet and lower part of legs are covered with scales which are homologous to the scales of reptiles. Claws have same basic characters in birds and reptiles. Besides, birds are covered by feathers which are homologous to the reptilian horny scales, as they have a similar origin and develop from similar germ-buds.

2. Endoskeleton

(a) *Skull*. In both, the skull is well-ossified, *diapsid* and *monocondylic*. i.e. with a single occipital condyle to articulate with vertebral column. Bones of cranium and palate are arranged on same general plan. Middle ear cavity contains a single rod-like *columella*.

(b) *Mandible*. Each ramus or half of the lower jaw is made of 5-6 bones. The lower jaw articulates with the skull by quadrate bones of upper jaw.

(c) *Vertebrae*. Epiphyses are entirely absent in both the groups, whereas present in mammals.

(d) *Girdles.* Coracoid is well developed in both the groups. It is single in birds and divided into a coracoid proper and precoracoid in reptiles. A precoracoid is also present in *Rhea.*

(e) *Limb-bones.* The ankle joint is *inter-tarsal* i.e. lying between the distal and proximal rows of tarsals.

(f) *Ribs.* The ribs are provided with uncinate processes in birds, and *Sphenodon* in reptiles.

3. Digestive system

In both, the alimentary canals ends in a cloaca. Gizzard is present in crocodiles and birds.

4. Respiratory system

Respiration is pulmonary or by lungs in both the groups. In birds characteristic air-sacs are present, in *Chameleon* air-sacs are also present. The trachea is made up of complete tracheal rings.

5. Circulatory system

A four chambered heart is present in crocodiles and birds. R.B.C's are oval and nucleated in both the groups.

6. Brain and sense organs

Extinct Cretaceous birds had small cerebral hemispheres as in reptiles. In both the groups external pinna or ear is absent and the cochlea of inner ear is not coiled. Many birds have a ring of minute bony plates known as *sclerotic ossicles* around the eye as in some extinct reptiles.

7. Urinogenital system

Kidneys are metanephric in both the cases. Cloaca is divided into three parts receiving rectal, renal and genital openings. In *Ostrich* a solid penis is present similar to that of *Chelonia* and crocodiles.

II. EMBRYOLOGICAL EVIDENCES

Following close similarities are seen in both the groups:

1. Internal fertilization, preceded by copulation.
2. Both are oviparous i.e. lay eggs. The eggs are *telolecithal* (large amount of yolk in eggs), and enclosed in a calcareous shell.
3. The segmentation or cleavage is *meroblastic* in both the groups.
4. Blastodermal *area opaca* and *area pellucida* are present.
5. In both, the foetal or extra embroynic membranes formed are *amnion, chorion, allantois* and *yolk sac.*
6. Minute gill-slits are found in the sides of neck in embryonic reptiles and birds but disappear in adults.

7. Hatching is similar in both the cases.
8. There is no larval stage.

III. PALAEONTOLOGICAL EVIDENCES

The fossil record of bird is rather incomplete and only a few fossil birds have so far been discovered. The fossilized ancient birds *Archaeopteryx* is the important intermediate form between reptiles and birds. It dates back to the late Jurassic period about 140 million years ago. It was discovered in a state quarry at Langenaltheim, Bavaria (Germany) in 1861 by *Wagner*. *A.lithographica* provides strong evidences of reptilian ancestry of birds.

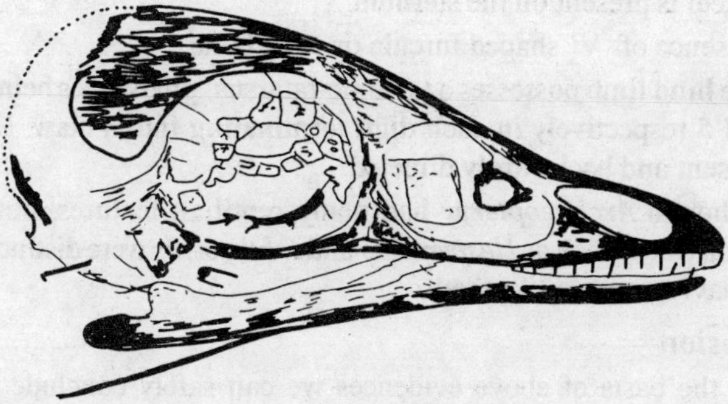

Fig. 13.1. *Archaeopteryx*. Skull.

Archaeopteryx was a small bird about the size of crow, but long lizard like tail-bearing feathers. It shows both reptilian and avian features.

A. Reptilian Characters of *Archaeopteryx*

1. Elongated lizard-like body with long jointed tail.
2. Skeletal bones are solid i.e. not pneumatic.
3. Strong jaws are provided with sharp homodont teeth.
4. Cervical vertabrae are fewer (9 or 10), than any bird.
5. Vertebral centra are *amphicoelous* as in *Sphenodon*.
6. Ribs single headed and without uncinate process.
7. Sternum is poorly developed.
8. Sclerotic ossicles are present around eyes.
9. Tail is composed of about 20 free caudal vertebrae.
10. Metacarpal bones are separate and carpo-metacarpus absent.

11. The three digits of the hand are clawed. The number of phalanges are 2, 3 and 4 in the first,second and third digits as in reptiles.
12. Scales are present on body and limbs.

B. Avian Characters of *Archaeopteryx*

1. Presence of feathers as body covering.
2. Fore limbs are modified into wings with only 3 digits each.
3. Tail bears elongated rectrices (tail feathers) arranged in two rows.
4. Brain box or cranium is large.
5. Elongated jaws are present in the form of beaks.
6. Scapulae are slender,curved bones.
7. A keel is present on the sternum.
8. Presence of 'V' shaped furcula or wish bone.
9. The hind limb possesses a tarsome-tatarsus, phallanges being 2, 3, 4 and 5 respectively in each digit terminating into a claw. Hallux is present and backwardly directed.

Although *Archaeopteryx* had many reptilian features, but extinct Cretaceous birds such as *Hesperornis* and *Ichthyornis* were distinctly birds except having reptile-like teeth.

Conclusion

On the basis of above evidences we can safely conclude that the birds and reptiles are very closely related and the birds have evolved from some reptilian stock.The ancestral birds which were more reptile-like underwent further modifications in the course of evolution due to new mode of aerial life.Thus flight adaptation may be regarded as the glorification of birds.Thus, *Huxley*, is justified in making the statement *'Birds are glorified reptiles'*.

14

Flight Adaptations in Birds

Adaptations can be defined as, "Morphological, anatomical and physiological changes or developments in the organisation of an animal which are suited to a particular environment are collectively called adaptations."

Birds are characteristically flying animals, and *Young* (1958) described them as *masters of the air*. There are several structural details with modifications. The following features may be enlisted as the important adaptations.

1. ***Shape.*** The shape of the body represents the sum of all its several adaptations. In birds the body is boat-shaped, which offers minimum resistance to air during flight.
2. ***Presence of feathers.*** The body of bird is covered over by smooth, closely fitted and backwardly directed contour feathers, which make the body streamlined and reduce the friction of air. They help in maintaining body temperature as they are of non-conducting nature.
3. ***Flight organs.*** A pair of wings forming flight organs which are modified fore-limbs beset with long quills and small covert feathers. These feathers form a broad continuous surface for striking the air in flight and supporting the bird in air.
4. ***Bipedal locomotion.*** For balancing and adjustment of the body, hind-limbs slightly move forward on the thorax and act as locomotory organs.
5. ***Short tail.*** The short muscular tail bears a series of long but light caudal feathers or rectrices arranged in a fan-like manner. They assist in steering, to sudden check flight and as a counter balance in perching.
6. ***Equilibrium.*** The wings are attached high up on the trunk, high position of light organs like lungs, position of heavy 'flight' muscles, and digestive organs lower the centre of gravity and preventing turning over of body with air.

7. ***Mobile neck and head.*** The neck of birds is very long and flexible. The beak helps in preening, feeding, nest building, offence and defence etc. For all these purposes movement of head is very important.
8. ***Perching.*** For arboreal life birds have very characteristic phenomenon of perching (holding of twig automatically at sitting by hind-limbs) with the help of pull on felxor tendons.
9. ***Integument.*** The loose skin is a modification for flight. It is responsible for extensive movement of the skeletal musculature.
10. ***Endoskeleton.*** It is light and provides large surface for attachment of muscles. It is built on "hollow girder principle" combine strength with lightness.

(a) The bones are pneumatic or hollow and due to pneumatic nature, weight of the body is reduced which is helpful in flight.

(b) The skull bones are fused and light. Teeth are absent in jaws.

(c) The first four thoracic vertebrae are fused to form a firm fulcrum for working of the wings.

(d) Last thoracic, lumbar, sacral and caudal vertebrae fuse to form *synsacrum*. It provides surface for the attachment of ilia. Pelvic girdle and synsacrum provide plate like out come to support the weight of body.

(e) The last four caudal vertebrae are fused to formed the *pygostyle* for the attachment of rectrices or tail feathers.

(f) Sternum is well developed to support the visceral organs, moreover a keel is present for the attachment of flight muscles. Keel is reduced or absent in flightless birds.

(g) Pectoral girdle is strong. Clavicle and inter-clavicles are fused to form 'V' shaped furcula which helps in flying. It is absent in flight-less birds.

(h) The fusion of distal tarsals with the metatarsals to form tarsometatarsus and that of proximal tarsals with the lower end of tibia to form tibiotarsus, help to strengthen the legs for bipedal gait.

(i) The uncinate processes of thoracic ribs help in producing compactness, necessary for flight by concentrating the mass.

11. ***Digestive system.*** The rate of metabolism is very high, food requirements are great and digestion is rapid. Crop is present for storage of food .Gizzard is specialized for crushing. Rectum is reduced and cloaca is capable of reabsorption of water from faeces and urine. Gall bladder usually absent.

12. ***Respiratory system.*** To meet the extensive rate of metabolism greater amount of oxygen is needed by the body tissue . Hence, respiratory system is highly developed. The lungs open into several air sacs. Air sacs are unique features of birds. These are thin walled, membranous and non-muscular. There are 9 air sacs in pigeon. Double respiration occurs, gaseous exchange is possible during inspiration and expiration due to presence of air sacs. There is no dead space.
13. ***Circulatory system.*** The heart is completely four-chambered and there is a double circulation. Large amount of haemoglobin is present in R.B.C's for rapid and perfect oxygenation, providing more energy. The body temperature ranges between 40-46°C.
14. ***Brain.*** Birds have acute sense of sight and intelligence so that optic lobes and cerebrum are well developed. Olfactory lobes are reduced. The cerebrum is also well developed.
15. ***Sense organs.*** The eyes are very large in relation to the head.There is an outer ring of sclerotic plates for resisting the pressure of air during flight. Pecten helps in lubrication of eyes and in detection of minute movements at great distance.
16 ***Excretory system.*** To reduce the body weight, the urinary bladder is absent.Water is reabsorbed in urinary tubules and cloaca. The urine is eliminated immediately in the form of uric acid and not retained for long period in the body.
17. ***Reproductive system.*** Only left ovary remains functional,the right being absent or vestigial. A single ovary also results in reducing the body weight, and the number of eggs production is also reduced.

Revision Questions

1. Describe various flight adaptations found in birds.
2. What is adaptation ? Describe how *Columba* or any other bird is adapted to its arial mods of life.

15

Migration in Birds

Migration (L. *migrare* = to travel) of birds is remarkable event of animals. For centuries the phenomenon of migration has been primarily associated with birds, indeed, judging by the references to the subject in the earlist literature, migration first observed in birds. Since civilization developed in a temperate region of the world where migratory movements are especially pronounces, it is hardly surprising that birds migration has long received attention. Man could not help noticing the flocks of birds in the spring and fall, the seasonal disappearance of some species and reappearance of others, nor could he help being interested in what he saw and eager to investigate what he could not understood.

Today there is an enormous literature on birds migration, based on extensive studies in Europe and North America. And yet the causes and process of migration are not fully understood. According to *Cohn* (1935) migartion can be defined as, "A periodic passing of animals from one place to another". As a rule, it is a response of an animal population to changes its environmental conditions.

Birds are more uniformly migratory than any other group of animals. Nearly all the orders of birds include species which perform migration ; in other vertebrates and in the lower groups of animals the migratory habit occurs in species scattered through a smaller proportion of orders.

Migratory and Resident birds

As earlier said birds are the greatest migrant, however, not all the species of birds participate in migration.

Bobwhite and ruffled sand grouse, do not migrate at all. Birds which remain throughout the year in a particular area are known as residents and others which actively participate in migratiton are migratory birds.

Kinds of Migration

Migartion in birds takes place as follows :

1. Latitudinal Migration

The latitudinal migration usually means the movement from *north* to *south* or *vice-versa*. Usually the birds migrate from their northern home to the southern home to pass the winter and with the advent of summer, again migrate to the north. For example several North American and Eurasian birds cross the equator to spend winter in deeper and warmer parts of South America and Africa. An opposite but lesser movement occurs in the southern hemisphere, where the seasons are reversed.

2. Altitudinal Migartion or Vertical Migartion

Some birds exhibit altitudinal migration i.e. they go to the mountain regions in the summer and return to the plains in winter seasons. Example: It occurs in the grebes and coots of Andes in Argentina, willow ptarmigan of Siberia etc.

3. Longitudinal Migration

Longitudinal migration takes place in the East-West direction. Thus starling moves from a breeding area in E. Europe towards the Atlantic coast to avoid continental water.

4. Irregular or Vagrant Migration

This migration is irregular and random. In some birds like herons, terns, etc., after breeding, the adults and the young may disperse in all the directions from their birth place.

5. Seasonal Migration

On the basis of the seasons, the birds migrate. There are following types of migration :

(a) *Summer visitors*. These brids migrate from warm southern to northern parts in spring season for breeding and nestling and return back in late summer or autumn. Examples. nightangles, swifts, swallows etc

(b) *Winter visitors*. These birds migrate southward and south-west in winter and go gack to North in spring. Example: snow buntings etc.

(c) *Partial migration*. In some cases, all the birds of a group of migratory birds do not leave the native land, but some birds stay on their breeding place. Such a migration is called partial magration. Examples: blue birds and many blue jays of Canada and Northern U.S.

(d) *Birds of passage*. Few birds during migration are seen for short time twice a year on their way to cold or warmer countries in spring and autumn. These birds are known as birds of passage. Examples . Sand pipers etc.

MODES OF FLIGHT IN MIGRATION

During their migration most migratory birds display following significant features.

1. Nocturnal and Diurnal Migration

Migration proceed by day or night. Many birds, ducks, geese, gulls, loons, shore birds and terns - travel by day or night, apparently indifferent to day light or darkness. But this is not the case with many other birds. Crows, eagles, falcons, hawks, herons, humming birds, swallows and swfits migrate only during day time while nearly all passerine birds except. swifts and few others, migrate primarily duimg the night from after sunsent until dawn.

For the nocturnal migration two explainations have been suggested:

(i) Movements of night affords birds, which normally live in thick vegetation and rarely take long flights away from it, the protection of darkness against their diurnal predators.

(ii) Movernent by night affords birds the opportunity of using all the day hours for feeding, thereby enabling them to built up sufficient energy resource for sustained long distance flights.

2. Range of migration

Extremes in distances traveled by migratory birds are represented on the one hand by high-mountain species, which merely pass up or down slopes of several hundred to a few thousand feet (Himalyan snow partidges cover a distance of one to two miles only while chicades come down nearly 8000 feet) and on other hand by Arctic tern (*Sterna paradisaea*). It makes the longest flight of any bird, migrating 11000 miles from the Arctic, where a part of the population breeds, to its wintering area adjacent to the pack ice around Antarctica. They return by the same route. Thus they cover 22000 miles in to and fro migration.

The golden plover holds the world record with the longest nonstop bird flight from Hudson Bay and Alaska to S. America, a distance of 24000 miles. It never stops for a rest and covers the distance in 48 hrs.

3. Altitude of migratory flight

The recent analysis of migratory flight by means of radar shows not only that birds move at altitudes averaging higher then formerely believed, but also that their altitude varies widely depending on the circumstances.

Nisbet (1963) studying radar lights of nocturanl fall migrants above Cape Cod, Massachusetts and the overlying oceans, found that most frequent hight was usually between 1500-2500 feet. About 90% of the birds, probably small passerines, were below 5,000 feet. On some nights they were lower than 2500 feet while on others they were upto 6000 feet or even 8000 feet.

Birds fly higher by night then by day. *Lack* (1960) first noted this tendency. Later *Eastuvod* and *Rider* (1965) proved that the tendency is significant. They showed that 80% of birds fly below 5000 feet at night and 80% below 3500 during day. They further demonstrated that migrating birds, in a 24 hour day, have a tendency to fly at the lowest altitudes in the afternoon and the highest just before midnight.

Direct observtions from aircraft proved that large birds can fly over the highest mountains ranges. The yellow-billed chough (*Pyrrhocorax graculus*) was actually found on Mount Everest at an elevation of 27000 feet. The use of radar in determining heights of migration shows that while most birds rarely exceed 8000 - 10,000 feet, a small proportion of migrants, particularly the stronger fliers, nonetheless attain great height, in some cases astonishing.

4. Velocity or Rate of migration travel

Most passerine birds fly at ground - speeds averaging 18-25 miles per hour. Stronger filers such as ducks, falcons, hawks, swifts etc. attain much greater speed. *Schnell* (1965) using doppler radar equipment measured the ground flight-speeds of 17 species birds in northern Michigan. He recorded on windless days two speeds of spotted sandpiper (*Actitis macularia*) at 25 miles per hour. Strong winds can significantly affect flight-speed. Hawks cruise along at 30-40 miles per hour, shore birds 40-50 miles per hour and ducks 50-60 miles per hour. *Stuart* recorded the greatest speed of two species of swifts which is 171-200 miles per hour. Besides strong winds, rains and hails also cut down the speed. Once the brids reached a higher altitude inland, their speed increased by about 10%.

5. Flocking during migration

During migration birds species show wide differences in flocking habit. A number of diurnal migrants notably hawks, kingfishers, swifts and other predators, are little inclined to move in group i.e. they travel solitarily. But majority of migrants, diurnal or nocturnal, exhibit the flocking habit.

Certain species migrate in flocks strictly of their own kind. These are usually birds whose flight speed, feeding habits or roosting preferences are so individual as to make them incompatible travelling companions. The common nighthawk (*Chordeiles minor*) and chimney swift (*Chaetura *pelagica*) are good examples of birds which migrate in their own companey.

Majority of species travelling in flocks, whether unmixed or mixed, give call notes. Unmixed or mixed flocks of some species of smaller birds such as certain sandpipers and plovers-fly in compact formations, all individuals in the flock simultaneously performing almost identical meneuvers. Majority of passerine birds travel in flocks that are loosely formed, though still cohesive, the individuals proceeding in the same direction. At night, cohesion of flocks is probably maintained by call notes. At the same time, the call notes serve to space out individuals in each flock so that they will not collide with one another. No flock of migrants appears to have a persisting leader. Whether or not flocks remains intact for the duration of migration has not been determined, but during migration flocking is undoubtedly advantageous to the individual concerned. V-shaped formation help to conserve energy by favourable air current for all the individuals in the flock except the leader. When frightend, the leader drops back and is replaced by another bird in the flock.

Unmixed or mixed flocks may contain only immature individuals, only adults or only adults of one sex, or they may certain individuals of all ages and both sexes. Flocks of geese and cranes may be comprises of one family or several families.

6. Regularity of migratory travel

Dispite of the weather effects on migration, the migratory travel over a period of years is regular on the average. If one keeps records for several years of the days when common summer resident species of birds first arrive in the spring, and computes average dates of arrival of each species, he can eventually predict within a few days when these species will appear.

There is much less regularity in the arrival and departure of transient species in the fall, thus the securring records is difficult. Through years of observation it has been found that in United States, at Washington barn swallows arrive approximately on twelth April, house wren on twelth April and chipping sparrows twenty second April.

7. Course of migration and migration routes

Species vary greatly in their course of migration and use routes. A few species move south over one route and return by another, a few species move eastward or westward before going south, and a few species have spectacularly long routes that take them as for south as southern south America.

Birds probably make use of several of the possible clues available for navigation in different parts of journey depending on the conditions. Sea routes are followed by marine birds. The coast route afford migration highways for a large number of migrants. River-valley routes are followed by those brids which migrate from plains to hills or vice-versa. Very rerely the birds cross mountain ranges. Sea routes, costal route, valleys and mountain ranges provide good land marks for the migrating birds, which enable the birds to recognize and remember the route.

Problems of Navigation

It has been observed that every year the offsprings of the migratory birds take the same route which their parents had taken. How they are directed on their path is still a problem. However, there are few explanations for it.

It is covenient to considered three types of orientation ability. Type I is navigation by visual or other land marks, Type II flying in a particular compass direction, and Type III is true navigation, the capacity to orientate to specific goal from an unfamillar area. This third type is rare and most migrants use combination of the first two.

Topographic clues can give long range guidence in daylight. Migrants fly high and at 2000 feet the horizen is 25 miles in either direction. Landmarks may be learned. Ducks and geese travelling in family groups transmit the information from generation to generation. Most birds little use of land marks except for the last stages of homing. They navigate mainly by the sun and stars. Pigeons fitted with translucent lenses that allow localization of the sun but not recognition of landmarks can home to the loft from up to 80 miles.

If landmarks are little used migration and homing must depend largely on type II navigation, by the sun or stars, and there is now much evidence of this. Displacement of starlings (*Sturnus*) during migration has shown that adults can alter course and arrive at their planned destination, where juveniles continue as before. The mechanism is evidently of type II. Such redirection would be useful to compensate for wind drift. But it is probable that much migration in fact makes use of the wind. The economy of energy moving down wind has obvious selective value and this has made birds into very good meteorologists. Migration occurs on a small proportion of days, when conditions are favourable. Thus autumn movements occur on the side of a high-pressure cell behind a cold front and spring flights on the west of a high pressure area ahead the cold front of a low. Birds often fly at the height where the wind is most favourable.

Karmer's experiment

The importance of orientation by the sun was first shown in experiments by *Kramer* (1951). Caged starlings show a restlessness at times when migration is due and flutter in the appropriate direction in relation to the sun. This would explain starting in the right direction but navigation on a long flight implies allowing for the sun's motion. *Kramer* showed that birds have a clock that does this. He trained to orientate in a particular compass direction by the morning sun and found they maintained this so long as they could see it or even a lamp bulb. If the bulb was kept stationary, however, they changed their angle as if it had moved. The bird's clock can be shifted by an artificial day/night cycle and starlings trianed to find food in a particular compass direction showed the expected shift of 15°/h which is the rate of sun's motion.

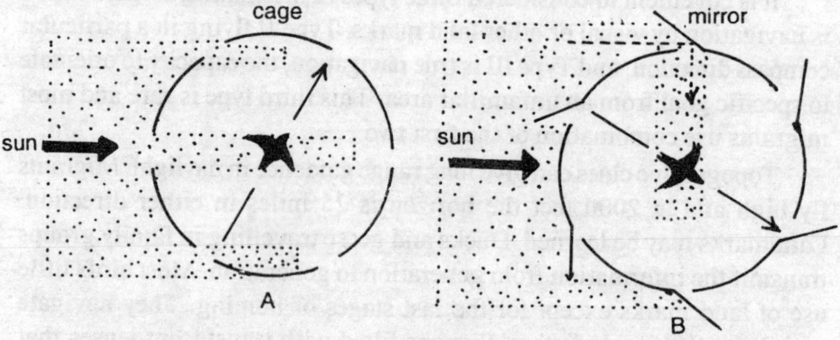

Fig. 15.1. Kramer's experiment with starlings to show that day migration use sun as a compass.

Such a sun compass can keep a bird on course but to set its course after displacement is a more difficult matter involving very delicate bicoordinate navigation. It has been satisfactorily proved that pigeons return home after release from a distant point even if they have never been there before, nor have had any previous training in returning from situations out of sight from the loft.

Matthews (1955) and others have shown that upon release the brids fly off towards home provided that they can see the sun. Capacity to navigate by the sun would depend upon determination of position on two coordinates by observations of the sun's altitude, azimuth, and/or movement. This would imply the use of a very accurate chronometer as a menas of determining the difference between home and local time. It is hard to believe that all this could be achieved in the few seconds following release, and probably only the azimuth is used with magnetic or other clues.

Nevigation by star light has been studied by following the direction of the restlessness of birds in a planetarium.

According to *Middendroff* and *Yeagler* the birds navigate through responses to the earth's magnetic field to the mechanical coriolies effect produced by the rotation of the earth. Particles of magnetite have been found in the brain, and muscles of the head and trunk of pigeons and sparrows. Magnetic force may influence the cells and /or receptors in muscles (*Presti* and *Pettigrew*, 1980).

According to some ornithologists homing instinct and telluric currents enable the migratory birds to definite routes.

In spite of the evidences for navigation by the sun and stars much remains uncertain. It is not clear what determines the initial compass bearing and in particular what is responsible for the reversal of migration between spring and autumn. These questions are especially acute in the case of birds that cross the equator. A further problem is that although birds usually avoid flying in clouds they seem to have some ability to maintain a course when forced to move under overcast skies or even in fog, though they may twist and loop along the way. Probably they use wind direction and other meteorological information but there is some evidence that birds can use magnetic clues.

CAUSES OF MIGRATION

What starts migration ? It is still an unsolved question. Various factors have been put forwarded by as the causes for migration. These are as follows :

1. Gonadal stimulus

Migrations are obviously correlated with the seasons and changes in day-length certainly play a central part in initiating them. It was first shown by *Rowan* in 1925 that juncos (*Junco*) or crows submitted to artificially increased day-length in winter show precocious migratory movements if released. The longer days also caused growth of the gonads (which regress in the autumn) and *Rowan* supposed that the secretion of sex hormones was the stimulus to spring migration and that its reduction produces, the autumn one.

However, studies with several migrant species have shown that artifically prolonged day length can produce hyperphagia, increased body weight, deposition of fat and restlessness even in the late autumn. Non-migrants showed no such effects. Maniputation of the photoperiod can be made to produce the tendency to move either north or south bunting (*Passerina*).

2. Scarcity of food

The increase of food availability in spring and its decrease in autumn may be actual triggers to the migration change. But this fail to explain as many brids are known to migrate even when there is planty of food.

3. Antipituitary hormone

Bissonnette (1931-37) concluded that decrease in pituitary activity in winter causes birds to migrate southward. During summer the pituitary hormone becomes active and gonads start developing. So, birds retun to north.

4. Thyroid hypothesis

The hormone secreted by thyroid gland regulates the metabolism, therefore, it also controls the migration.

5. Metabolic stimulus

Walfson (1942) concluded that increased day length stimulates deposition of fat. The large deposition of fat is believed to provide the stimulus for migration.

ORIGIN OF MIGRATION

Several theories have been put forward to explain the origin of migration. The important theories are :

1. Northern home theory

According to this theory, the original home of the birds was in the Northern hemisphere. In the Pleistocene period glaciation took place in North, the birds were forced to South. When glaciation receded the birds returned to their original home i.e. North. In next winter they again forced to South retuned back during favorable conditions to North. This is believed to have originated a migratory rhythm which in due course of time become hereditary. This theory fails to explain migration in those parts of the world which were never affected by glaciation.

2. Southern home theory

This theory assumes that the ancestral home of the birds was in the South. Due to over population, some birds migrate to the North during breeding season and returned back as soon as nesting was over. This theory still needs confirmation and cannot explain all types of migration satisfactorily.

Significance of Migration

Migration is useful as well harmful to the birds. Advantages of migration are following

1. Birds get suitable place for breeding.
2. The birds get plenty of food supply in the place of migration.
3. As a result of arrival at warm places from cold places, the birds get longer day time to search their food.
4. As a result of migration, the birds escape cold and stormy or excessive heat.

Disadvantages of Migration

Migration is dangerous for all birds. In their long flights over land or water they are likely to meet disaster through vagaries of the weather. While migration are adjusted to the normal alternation of fair and inclement weather during spring and fall, sudden and unseasonable changes in the weather occasionally have serious effects. After the storm, one observer reported an estimated 5000 dead birds washed up on a one-mile stretch of shore.

Man has created awesome hazards for migrating birds by erecting light-houses with strong light beams and illuminating various tall structures. During nights in the spring and fall when migration proceeding at a low elevation because of an unsurmountable cloud layer, passing birds are attracted by brightness and, approaching it, soon become blinded and fly into its source, killing themselves.

Airport ceilometers indirectly cause mortality among small, nocturnal migrants. A direct cause of mortality among small nocturnal migrants are television towers erected to hight of 900-1000 feet or more.

Revision Questions

1. Write an essay on migration of birds.
2. Write an essay on migration in birds explaining the latest theories for the cause of migration.

16

Flightless Birds or Ratitae

Ratitae are the flightless birds with reduced wings and no sternal keel. They have long legs and curly feathers. It has been suggested that they diverged so early from the ancestral avian stock, that they never passed through a flying stage. *Romer* regards the acquisition of flight in birds as an adaptation for escape from enemies on the ground. If this view be justified we can understand why some descendants of flying birds became flightless. Thus we see that the Ratitae have stem from flying ancestors and have evidently lost their ability to fly as their size increased. This has led to considerable change in the bones, muscles, and plumage.

CHARACTERS OF RATITAE

The Ratitae birds differ from all other birds in the following respects :

(i) They are almost flightless and usually of large size.
(ii) The sternum is devoid of keel.
(iii) Wish bone or furcula absent.
(iv) The long axis of the adjacent parts of the scapula and coracoid are nearly parallel.
(v) The posterior end of the palatine and the anterior end of pterygoid are not at all articulated with the basisphenoidal rostrum forming *desmognathus* type of palate.
(vi) The feathers are 'primitive' in that they have no hooked barbules, so that the barbs are free.
(vii) There is no syrinx.
(viii) The wings are reduced in size.
(ix) There is no pygostyle.
(x) Oil glands are absent.
(xi) The uncinnate processes are small or absent.
(xii) The plumage covers the whole body with little uniformity expect that there are naked places on the head, neck and the extremity of

the abdomen.

(xiii) The young are precocious or nidifugous.

(xiv) Strong legs with leg bones without air spaces.

The flightless birds such as Ostrich, Cassowary, Kiwi, etc., with reduced wings and have no keel. Long legs with curly feathers placed them in a distinct group and suggested as primitive. Indeed, it has been suggested that they diverged so early from the ancestral avian stock that they never passed through the flying stage. The substantial cerebellum is an indication that they are descended from volant ancestors and represent several different evolutionary lines.

ORDERS OF RATITAE

Order-1. Struthioniformes (Camel birds or Ostriches)

These are the largest of living birds and are confined to the plains and deserts of Africa and S.W. Asia. They are highly specialized as runners than are any others. African Ostrich attains a height of 2.5 meters and about 150 kg. in weight. Male ostrich is the largest living bird. It takes about 4.5 meter strides and can travel 65 km per hour. When the animal is at full speed the wings are stretched out as balancers and probably partially lift the weight of the ground after the manner of the hypothetical pro-avian cursorial ancestor of the birds. The wings are reduced and the barbs

Fig. 16.1. The Ostrich.

are not attached to one another. The wing has 16 primaries, 4 alular and 20-30 secondary feathers. The flight muscles are not well developed. The legs are strongly developed; naked and pinkish. The tarsus in sexually mature male has red horn plates, while in sexually mature female they are black. There are only two toes per foot and these are covered by broad pads for running on sand. The foot is very much like that of the camel. Head is small and has large eyes. It is mounted over a long pink skinned neck. The enormous size of rectum and ceaca is a unique character.

Ostrich is gregareous bird and travels in groups of 3-20. It is polygamous and during breeding seasons one male looks after four or five females which lay eggs in a common nest. Male possesses a penis and copulation is preceeded by some sexual display by the male, in which it drops to its heels before a female. The nest consists only of a pit in the sand. The eggs, depending on the age of the hen, are 127 x 103 mm to 160 x 130 mm. The three pound olive green egg is incubated by the female. Brooding of eggs is necessary, for the egg would be chilled and doubtless killed by the low noctunal temperature, characteristics of deserts and arid regions. The bird is omnivorous and its feathers have ornamental value.

Order-2. Rheiformes (Rhea)

The rhea occupies the same ecological position in South America as the ostriches in Africa. The rhea is much like the ostrich in general apperance and in habits. It is smaller and less specialized for running than the African Ostrich. Standing up right they reach 1.70 meters. They may weight upto 25 kg. There are three forwardly directed toes with heavy claws and pads. The hind toe is absent. The wings are better developed

Fig. 16.2. The Rhea.

and the feathersless plume like that in the Ostrich. The plumage consists of gray and rufous feathers. There are large scales on feet. The head, neck and thighs are feathered. The gut and caeca are large. They are grass and leaf eaters. The male excavates a shallow nest in the ground and surrounds with few leaves and grasses. They have a well developed syrinx. Numerous eggs are laid in the nest. The rheas are swift runners. The palate is desmognathus. The genus *Rhea* also existed in Pliocene.

The urine is stored in an expansion of the cloaca and is eliminated in liquid form. Male with copulatory organ which is extrudable. The eggs measure 135 x 95 mm and weigh 530 to 680 gms. The eggs are elliptical with a shiny surface and are ivory or golden yellow with black stripe-shaped processes. The young at first are yellow with black stripes on the back; after two years they resemble their parents.

Order-3. Casuariformes (Emu and Cassowaries)

These have compressed beak usually with a helmet shaped bony knob on the head. The neck and body are densely feathered. Feathers have aftershaft which is almost equal to the main shaft. The palate is desmognathus. The vomer is large and articulates with palatine and pterygoids. The palatal bones and sphenoid touch one another. The wings are very small. There are three toes on each foot. The legs proportionately shorter than in the two preceding groups. Emu (*Dromais*), and cassowary (*Casuarius*) of Australia, New Guinea and East Indis are kept under this order.

Fig. 16.3. The Emu.

Emu inhabits the open grassy plains. It is nearly 6 feet tall. Next the Ostrich, Emu is the longest bird. It runs fast. Inner toe of each foot bears a powerful claw. This can rip the belly of a dog. It is herbivorous in diet. It is purely monogamous differing in this respect from other Ratitae birds.

The male incubates the eggs that are laid to the number of a dozen or more in a hollow scrapped out of the surface soil. The flesh is palatable and the sub-cutaneous fat is used by the natives for oil.

Fig. 16.4. The Cassowary.

Cassowaries live in the forests of Australia, New Guinea and neighbouring islands. They are about 150 cm tall. They are fruit eaters and are confined to the densest part of the forest. The plumage is much like long, soft fur and is used for weaving rugs and ornaments. There is a well marked blue head with a helmet like black shield or casque with green sides. The casques perhaps help the bird in pushing its way through the thick vegetation and also in the camouflage. Neck is short and foot with three toes, the inner bears a long sharp claw for defence. The bird is nocturnal and frugivorous.

Order-4. Apterygiformes (Kiwi)

Kiwi is the small running or fligtless bird. It is a national bird of New Zeland. It has a size of a common fowl. The wings are vestigial, hidden and non functional. It is without any keel. It has simple hair like or bristle-like feathers. There is a long bill with terminal nostrils. The tail feathers are also absent. There is no clavicle. The short powerful legs are covered with scales. The three anteriorly directed toes are equiped with claw and meet the ground. The fourth toe is small, directed backwards and raised above the ground. the bird is swift and noiseless in movement. The palate is desmognathus. It shows large basipterygoid processes. The eyes are remarkable for their small size. The kiwis are nocturnal and remain concealed in burrows of their own making during the day. The kiwi feeds mainly on insects and earthworms. The bird seizes the worm with the long beak and gently pulls it out of its hole. The sense of smell and the parts of the brain related to it are better developed than in other birds.

Fig. 16.5. Kiwi.

The male has a penis and female lays fertilized and strickingly large eggs in the burrow or nest in the earth. There is a controversy regarding the incubation by eithr sex.

Order-5. Dinornithiformes (Moas)

It also includes flightless terrestrial birds. The group is represented by gigantic moas (*Dinornis*) which was restricted to New Zealand where it became extinct. Some forms attained an enormous size upto 10 feet. About a century ago the skeletons of gigantic wingless birds were found

Fig. 16.6. Moas.

scattered about the plains. In most species the wings and the pectoral bones have disappeared completely, and there was no trace of keel. The beaks were short and the massive legs have four toes. They were probably herbivorous. The head and neck much like those of Ostrich.

Order-6. Aepyornithiformes (Elephant bird)

It was represented by the giant flightless Madagascar *Aepyornis,* the elephant bird. The species, existed in the Pleistocene. Some were larger than Ostriches, with eggs estimated to weight more than 10 kg, presumably the largest single cell, that have existed. They possessed relatively tiny wings powerful legs with four toes. No doubt this fine bird was hunted out of existence by the native tribes of Madagascar. Possibly the collecting of their eggs was more destructive to the species than was the slaughter of adults.

Order-7. Tinamiformes (Tinamus)

These birds have traditionally being placed in the position given presently, which otherwise can be placed with near *Galliformes* of the super order *Neognathae.* They are terresterial birds rather like hens and are essentially cursorial but they can fly over short distances due to the presence of short and rounded wings. The sternum is keeled. They have primitive type of palate characteristics of all the Ratitae. The tail feathers are very much reduced. There is but one family i.e. Tinamidae of more than 50 species which are confined to the South Mexico and Central and South America. When flushed *Tinamus* usually run rapidly and may conceal themselves, they do not perch and are essentially vegetarian. The

Fig. 16.7. Tinamus.

eggs are laid in the nest on the ground. The eggs are highly glossy. Small pygostyle is also present.

ORIGIN OF RATITAE

The origin of flightless birds still keep ornithologists at an uncertain position. It is still debatable point whether they had descended from cursorial or from volant flying forms. *Romer* held his view that these Ratitae had descended from flying type and the flying is an adaptation to escape from enemies and odd environment. If there are no enemies, birds need not to fly and this condition gradually brought a flightless condition. His hypothesis still owing to the distribution of Ratitae birds. New Zealand gets a passive distribution of Kiwis but there are no carnivorous mammals, indanger to them. Madagascar, South America, Australis, and Malayan Archepilago are rich is ratitae birds, but lack the carnivorous mammals. The other school of throghts says that they are not derived from a single ancestral stock, but independently from different ancestral stock or groups of birds. This seems hardly likely in view of the following facts. All of the Ratitae have retained the primitive type of palate. In Ratitae the males rather then females incubate the eggs. It is quite unprobable that this combination of peculiar features has developed independently in a number of separate ancestral stocks. Thus the origin of Ratitae is yet an unsolved problem. Probably they have evolved from a tertiary group of birds, that became wide spread over the world before the carnivorous placentral mammals made their appearance. They are absent from most of the continents due to the presence of carnivores.

17

Palate in Birds

Palate is the roof of vertebrate mouth cavity. The roof of the mouth cavity in Mammals is not at all homologous with that of other Vertebrates specially of birds. A new (false) palate is developed beneath the original palate by bony shelves projecting inwards from the bones of upper jaw. Thus the nasal cavity is extended and its internal opening which is placed right back in the throat. The statement that "the avian orders are based on internal features such as palate and arrangement of bones in skull" was given by *George J. Wallace* in his book "An Introduction to Ornithology." *I. H. Huxley* proposed four major groups of carinatal birds on the basis of their palates . The isolated Emu and other Struthian birds under Ratitae a separate division.

Palate can be defined as the combined nature of vomer, palatine, pterygoid, and basisphenoid in roof of buccal cavity. Basically it keeps the mouth and nasal passage quite apart from one another. In desmognathus birds like swans, the nostrils have anterior most position at tips of long slender bill and the plate in these cases is said to be the most complete one. The effect of extensive bony palate is to bring internal nostril back into proximity with glottis (*McDowell*).

CHARACTERISTICS OF AVIAN SKULL

In birds the brain case is always arched and spaceous and the bones are usually fused together and the surfaces are obliterated. The orbits are generally very large and there is a well marked interorbital septum, and the facial parts are prolonged into a beak consisting chiefly of the premaxilla. A rod like quadrato-jugal reaching behind to the quadrate.

The supratemporals are usually incomplete but the squamosal sends forward a process which joins the post orbital process of the frontals and there are no pre- and post-frontals. The orbits are not closed posteriorly from the temporal foassa, though in parrots the postorbital process of the lacrymal beneath the orbit. The squamosal is closely applied to the skull these features are all commonly found in birds, however, there are many differences in connection with the bones of palate type.

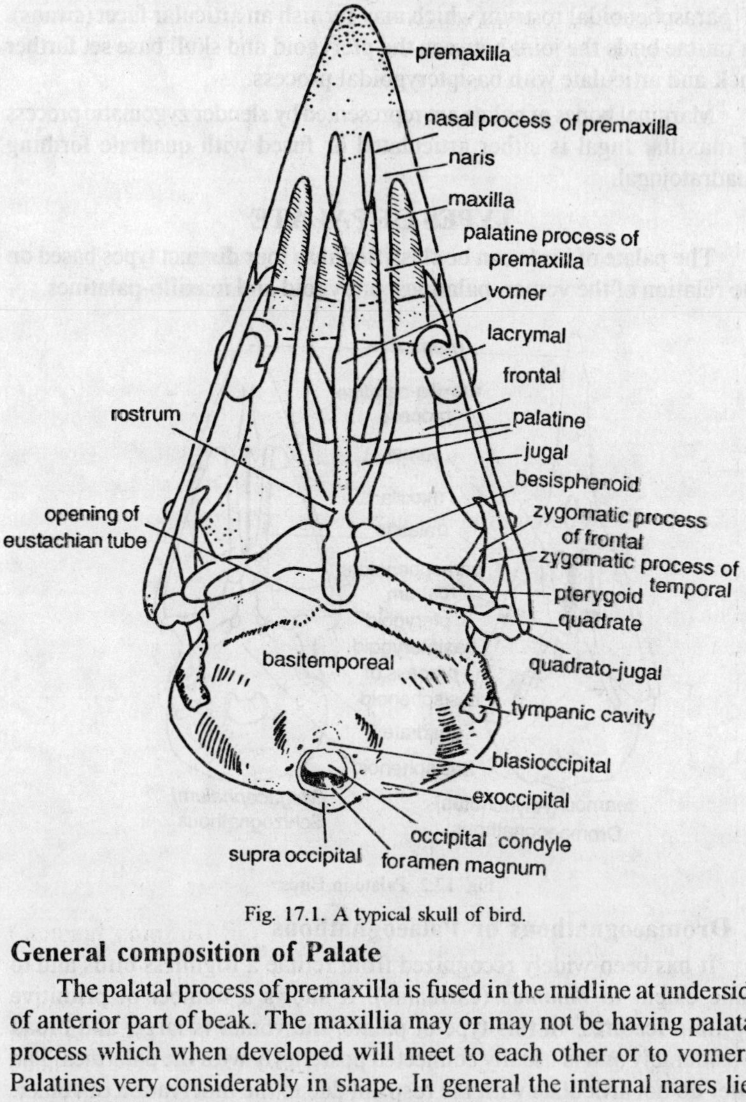

Fig. 17.1. A typical skull of bird.

General composition of Palate

The palatal process of premaxilla is fused in the midline at underside of anterior part of beak. The maxillia may or may not be having palatal process which when developed will meet to each other or to vomers. Palatines very considerably in shape. In general the internal nares lies surrounded by process of maxilla, vomer and palatine. A little posterior side of palate in mid line. There is a bone attached or fused above and lower edge of the interorbital septum. This is the parasphenoid. At its back the bone widens out into a broad basisphenoid. Pterygoid runs backward, and outwards from posterior end of palatine and meets to quadrate. In carinatae birds the pterygoid is usually connected with side of parasphenoidal rostrum which may furnish an articular facet (swans). In ratitae birds the joint between the pterygoid and skull base set farther back and articulate with basipterygoidal process.

Marginal bones of palate are represented by slender zygomatic process of maxilla. Jugal is either articulated or fused with quadrate forming quadratojugal.

TYPES OF PALATE

The palate of birds can be classified into four distinct types based on the relation of the vomer, palatines, pterygoid and maxillo-palatines.

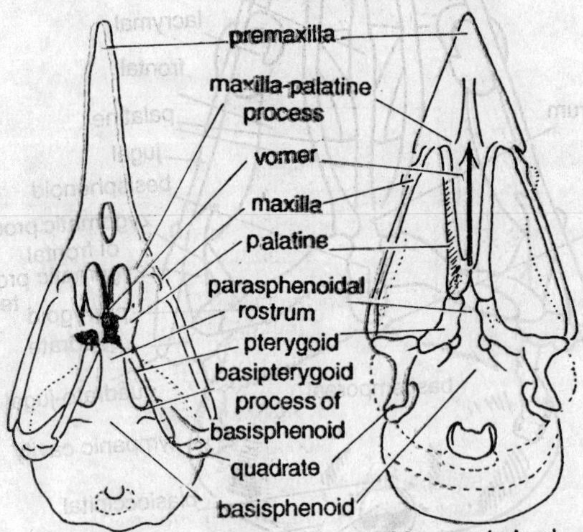

Fig. 17.2. Palate in Birds.

1. Dromaeognathous or Palaeognathous

It has been widely recognized from retitae a flightless birds and to some extent to *Tinamus* (carinatae). It shows a number of primitive reptilian features. In this type of palate, the vomer is large, and broad considerably and is usually connected posteriorly with the palatines. The latter do not articulate with the rostrum due to the innervation of vomer. The maxillo-palatine processes are comparatively small and do not unite with one another or with the vomer. The basipteryoid process of the basisphenoid are large and articulate with the posterior part of the pterygoids. In *Struthio,* the vomer is short and does not articulate with the maxillo-palatine. The palatines extend far beyond the maxillo-palatines. In *Rhea* the vomer is long and articulates with the palatine and pterygoid but does not join the maxillo-palatines. The short palatines unite with the maxillo-palatine. In *Casuarius* and *Dromaeus,* the long vomer articulates with the palatine. In *Apteryx,* the vomer is long and unites with the palatine and pterygoid behind.

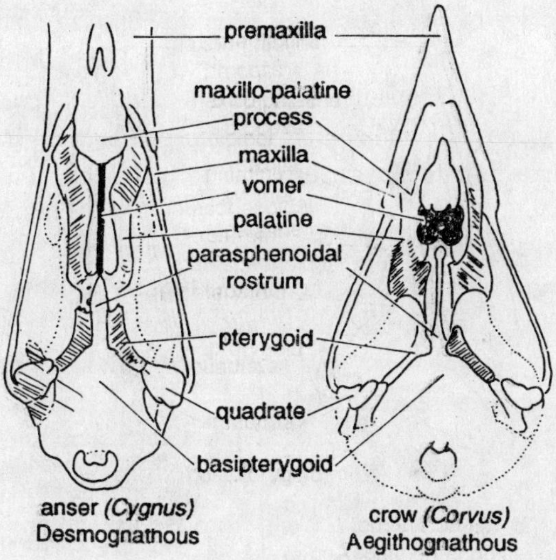

Fig. 17.3.

Schizognathous

In this type the vomer is either absent or small and pointed separately the maxillo-palatines. The basipterygoid processes are either absent or springing from the base of rostrum. The palatines and pterygoids articulate with the rostrum at the point where they join one another. The maxillo-palatine processes are poorly developed and not unite with each other with the small vomer, if present. This type of palate occurs in many carinatae birds like pigeons, fowls, gulls, plovers etc.

Aegithognathous

This is rather to say a little modified schizognathous and is showing the similar arrangement of bones except the vomer which is broad and truncate infront, instead of being pointed. Ex. Passers and swifts.

Desmognathous

In this type of palate the vomer is usually abortive or so small that is disappears in the skeleton. When present, it is always slender and tapers to a point in front. The maxillo-palatine processes of the maxillae are large and spongy and unite with each other in the middle line to form a flat spongy palate. Ex. duck, geese, storks, parrots and birds of prey.

18

Territorial Behaviour in Birds

The history of "Territorial behaviour in Birds" began as far as back as 1822 when it was found that birds defended plots of ground in the breeding season against members of their own species. Since then many wild guesses and hypothesis were put forth explaining territorial behaviour in birds and it was in 1903 that a well founded theory could be propounded to explain the phenomenon. Earlier many attempts, like that of *Bernard Altum* (1868), were made; although, the work of this German zoologist remained restricted to Germany and Ornithologists outside the country could not know of his work. *Eliot Howard,* quite unaware of Altum's work; published a small book on "*Territory in bird life"* which could attract the ornithologists every where on the earth; and the territory concept quickly became a major stimulating and guiding factor in field work on breeding birds.

DEFINITION OF TERRITORY

Noble has defined territory as "any defended area" and *Nice* (1914) accepted this as most satisfactory definition. The instinct of defending certain plot on the ground is specially developed in males, who guard the area at any cost. In words of *Gilbert White*—"during the amorous season, such a jealousy prevails between male birds that they can hardly bear to be together in the same hedge or field."

The modern definitions of territory are as below :

1. "Whenever sexual fighting is confined to a restricted are that are is a territory". *Tinbergen.*
2. "Isolated area defended by one individual of a species or by breeding pair against intruders of same species and in which the owner of territory makes itself conspicuous." *Lack*

For the study of territorial behaviour in birds a passerine or perching land bird is generally taken as an standard one, although on one species exhibits the examplary activities of this phenomenon. The phnomenon may be described under the following heads :

(1) Taking up of territory, (2) Staking the claim i.e. advertisement of territory possession, (3) Size and shape of territory, (4) Nesting and feeding ground, (5) Courtshio and display.

1. Taking up of territory

The territories may be taken up by males e.g. sea birds and some arctic wanders that pair upon migration, before they have arrived at the breeding place or by females e.g. phalaropes and button quails, however females have bright plumage and aggressive displays. Territories may be taken up by both the sexes altogether.

Winter territory

The Robins may inhabit only a few acres during whole of its life, though certain females wander and a few are even migratory. In winter both male and female hold individual territories, which together are almost exactly the size of the summer ones. There they defend like most of aggressive birds by song and dance. In spring the females stop their activity and very often mate with a neighbour rival male. In early autumn the territorial activity is low. The advantage of winter territory is that it keeps the species evenly spread in regions of scanty food and in small units invisible to predators than in large visible ones.

In most passerine birds, at the end of winter in early March the females and males share foot etc. with co-operation in a flock but soon quarrels begin its break between males of same flock, whch may not necessarily be over food. As the time passes the quarrels tend to get slightly more serious and prolonged and are accompanied by little exposition of breeding songs. The males pay short visits to nearby suitable places of their summer breeding grounds. These visits to small areas of breeding habitat lead to even more pronounced excursions and as the time goes on the period spend away from flock gets longer and longer and the area visited smaller and smaller until a day comes when male has completely left flocks and sleeps in what may now be called as "Rough Copy" of his breeding territory.

The male now goes less far in the field for foragging and feels entirely in loose area of his own territory. This area does not say loose and undefined for a long period but becomes restricted and well fought for against the other intruding males. It has a definite head quarter where male sings and displays.

It is now the female turn to leave the flock. When one of the females happens to pass through a male territory, the latter sings and performes dances to attract her. It thus convinced the female and male both now defend the territory by singing, dancing and fighting. The sexual chases became established as well and are pronounced enough when the couple has spent a few days together.

2. Staking the territory

To advertise the possession of a certain territory the birds use songs, chasing and fighting. By performing such activities, the birds make it obvious to others that they are in possession of that particular area. Such an exhibition may be explained under following sub headways :

Songs and call notes

The voice is merely a means of communicating their feelings. Singing is a natural expression of emotional intensity. It is best and solely developed in males. The song of male stops as long as his own female in the territory, but in case of any astray of female, he again starts singing to call her back.

Saryard fowl calls her chick in different notes. The sound produced for food, disaster and brood differ very much. The sound production is undoubtedly a secondary sexual character and has been developed as a means of bringing about the sexes together and of actual species recognition. Most of the birds have words expressing pleasure, pain, sense of danger and presence of food etc.

Fighting

The female helps the male in guarding the territory. Generally the song of a bird is primarily an announcement to the females of the presence of male and challenge to other males of his species, to keep them off this territory he is guarding. The fighting behaviour is best illustrated by taking an example of song sparrow by *Allen*. A certain song sparrow decided to have a nest in an adjoining hedge and had been announcing the fact of his cheerful song for several days. One unfortunate day he entered into it and discovered his own reflection in the window. He thought that it was another song sparrow trespassing his territory and with a hurry he dashed into the glass to drive its way. But there were so many glasses or windows with the result that he went without a mate from earlier April.

Chasing

Swans will chase any large animal from their territory, although ringed plovers will chase skylarks and linnerets; little ringed plovers will even try to chase sheep. We can perhaps explain this by pointing out that these birds are guarding the nest; ringed-plovers with their nest on ground have it exposed to the danger of being trodden on by mistake as well as deliberately raided. We can thus reach the conclusion that birds normally attack other species in their territory only when these animals tend to interfere with the breeding cycle.

3. Size and shape of territory

The extent of territory is not definitely fixed in all birds. With birds of prey it does not appear at all to be rigid, with others like crow and Dartford warblers all territorialism rather breaks down when population is high; and in other the later we may find the strange spectacle of several pairs sharing the same territory. But the constant minimum of half acre gives us clue. As previously hinted territories are not indefinately compressible or expansible. As Professor *Julian Huxley* says "Territories are like elastic disc, of which there is a lower as well as a higher number which can be placed together to cover a given area.

4. Nesting and feeding grounds

The nest building is entirely due to fidgetting whch may take various forms including making a scoop in the soil and pressing up pieces of grass after copulation. The place of nesting is chose by male and sometimes male brings the material and female uses it.

Regarding the feeding grounds, *Howard* was first to suggest that system of territory holding might confer advantage on birds by assuring them a regular food supply. Warblers, buntings etc. seek food supply within their territories. In territorialism ensures an even distribution of birds over an area, thus not stretching unduely the limits of food supply in any one place while neglecting sources of food in another.

5. Courtship and display

According to *Dr. Tinbergen* the male sings specially when his female is absent; his song increases when she refuses to copulate and it almost reaches its original strength; when the female begins incubation. This probably indicates that sexual potency still exists. The breeding of birds is nearly always seasonal even in torpics where the conditions are unifrom throughout the year. The changes in behaviour with reference to the gonads vary greatly with the species. Birds which have been social throughout the winter e.g. buntings begin to leave flock and assume territories and into which they invite the female, to copulate with her. Song is one feature of elaborate business of display and courtship. In simplest cases the display serves to bring the sexes together, to enable recognition and at a later stage as stimulus to copulation.

Since the male bird cannot grasp the female, copulation can only occur with her full co-operation. The male and female birds fly round and round their territory; male singing and displaying of the sexual desire. At a later stage the copulation occurs. The female now becomes restless and makes frequent visits to the nests which is completed within 5-15 days of first mating. The male on his part resumes singing as if he was unmated, however, female does not allow him copulate. The ultimate aim of courtship and display is copulation followed by nest building and then ovulation.

6. Breeding and care of young

The males take no part of incubation but continues his territorial song and aggressiveness and goes at intervals for foragging bringing food to the female, on the nest. After a fortnight the eggs hatch, young ones like most birds are at first naked and blind. They are helpless and are to be fed by the parents for a considerable time.

Young birds at hatching are of two kinds-precocial actricial. The former are those which resemble young chicken in that they are wide awake, covered down feathers and are able to follow their parents in search of food soon after drying. The actricial youngs are naked and helpless and are cared for a long time.

For a day or two the young ones are fed by the secretion of crops, after this by large quantities of insects. The young crow requires about 1/2 of its body weight the food daily. The number of trips made by female to the nest for feeding vary from sepcies to species. Humming birds feed their young ones in 20 minutes; House Wren 1217 times in 15 hours and 45 minutes. Often a group of parents goes in search of food altogether and at this time no aggressive behaviour can be observed.

The parent bird swallows food and give it to young in a partially digested state. Either the parent may put the food into nestlings mouth. The feeding of nestlings is not cyclic but one young is fed until his stomach is full, and then is fed the one who raises his neck high and mouth wide open. The youngs grow more and more noisy and active. After a week or 10 days the young begin to feed themselves. After another week they are no longer fed by the parents. In some cases the parents make the young birds to learn how to prey while on wings. In hawks this phenomenon is well developed, where parent with some food flies and makes the young one to fly after it and catch the food.

After some more days parents leave their children alone, lose the sense of territory and their different voice notes are replaced by ordinary flocking songs. In the following autumn the young birds of fresh flocks, and if migrating, start migration.

SIGNIFICANCE OF TERRITORIAL BEHAVIOUR

Territoralism among birds is a subject of much speculation and discussions are held on what the advantages do the birds get as a result of this phenomenon. A few uses are certain as enumerated below :

1. Food

Howard and his early disciples emphasised that the importance of ensuring an adequate food supply for the young lies chiefly connected with this behaviour of birds. Modern studies have greatly reduced our general belief in importance of this interpretation.

2. Pair bond

Territorial behaviour is surely important in many species in the formation and maintainence of bond between pair of birds. Since the two birds remain within a restricted area, they are able to recognise each other more readily than they could do in flock.

3. Reduction of interference

The reduction of interference of species in the reproductive activties is certainly important. In most territorial species the chances of fatal hybridization are reduced and a pure race or species remains pure.

4. Regulation of population density

Population of a species in a favourable habitat is one of the results of territorial behaviour. The over crowding is reduced and quarrels between individuals are reduced to minimum.

5. Reduction in less through predators

In normal flocks a number of birds are killed by predatory birds etc. Thus causing a considerable damage but during breeding when utmost safety and security is necessary; the birds disperse, in pair thus reduce the chances of being preyed upon since they are not now as conspicous as were in the flocks.

6. Disease resistance

Although this has not much evidence, the dispersion upto some extent stops the spreading of infections diseases among birds.

19

Beaks and Claws in Birds

The quality, we define, as life is perhaps more fully represented in birds than in other vertebrates or indeed in any animal what so ever. Following out the definition of the life of a species as the sum total of activities by which that particular type of organization is preserved, we shall find that the birds have many and very varied activities, by means of which a great deal of matter is collected into the bird type of organization. The birds are adapted for moving in the air, the most difficult medium of all.

The flight is the course characteristic that gives us most fully the feeling that the birds are active animals. One of the most striking features of bird life is that although the basic organization remains fairly constant, different types show a great variety of special features fitting them for numerous habits and habitats. Besides differing in behaviour in body form and in power of flight there are other characters in the shape of the bill and hence of food habits and in the details of many parts such as the feet, that makes fascinating studies in adaptation to environment.

No where in their radiation have birds diverged more widely than in their adaptation for obtaining food. Although the earliest birds probably ate insects and other small animals, many modern forms are specialised for obtaining particular kind of food from both animals and plants.

The modification of the fore limbs as organs of flight prevents their use in grasping and holding food, hence the bill and to a lesser extent the hind limbs must take over these functional modification for food getting.

The bill or beak is composed of an upper jaw or maxilla and an lower jaw or mandible From the figure (see skelton of fowl) it will be seen that *maxilla* is not strictly the whole upper portion and though the term is thus for convinence as is the pleural *mandibles* for the two jaws when mentioned simultaneously. The *rhamphotheca* (horny sheath) may be simple (undivided) or compound that is made up of several distinct pieces.

TYPES OF BEAKS

In birds the beak is the modified jaw bones without teeth. In different birds the beaks are modified for various functioning thereby show the adaptive radiation. The beak is mainly related to the feeding behaviour of the bird. Some common modifications are given in the following lines :

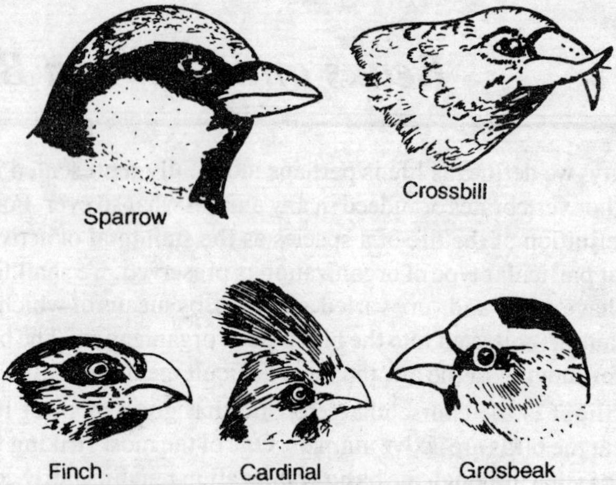

Fig. 19.1. Kinds of seed eating beaks.

1. Seed eating beaks

This type of beak is found in small granivorous birds like sparrows and finches. The beak is short, conical and stout. The birds pick up the grains which may partly be crushed. In crossbills is constructed in an unusual manner and it enables the bird to obtain seeds that lie deeply burried among the projecting scales of caves of pine.

2. Cutting beaks

The crow possesses a long, sharp beak with cutting edges. It is used for various purposes like picking, cutting etc.

3. Fruit eating beaks

In some birds like parrot, the beak is sharp, deeply hooked, massive and extremely strong. In such birds the upper beak moves slowly. It is adapted for gnawing or beaking open hard seeds and nuts. The bird like parrot uses it feet in holding the food while feeding. In hornbills beak is heavy and gratly developed than parrot.

4. Insectivorous beaks

There are many birds that feed upon insects. To catch the insects while on the wings, they develop comparatively small, wide, soft or strong beaks. The beak is notched at the tip in flycatcher but in night-jars the gap is deeply cleft. In swallows and swifts the beak is small, wide and delicate whereas in warblers it is long and slender. In tree creepers it is quite long and curved to take the insects out of crevices of tree barks. The beak is long, slender and slightly curved to turn the leaves as in hoopoe.

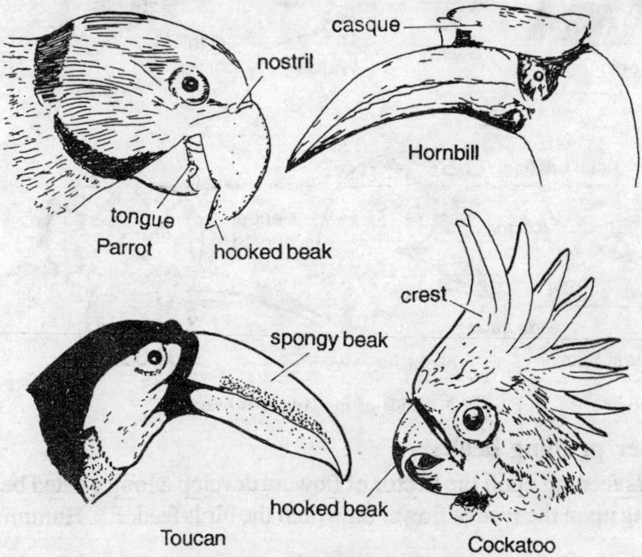

Fig. 19.2. Kinds of fruit eating beaks.

5. Wood chiselling beaks

In some birds the beak is elongated, straight and stout. It is used for drilling into woods to take the insects out and far the construction of nest. Ex. Wood peckers.

6. Mud probing beaks

In some birds the beak is elongated, slender and strongly built. It is used as a probe for thrusting the mud in search of worms. Ex. Sand piper, Jacana etc.

7. Tearing and piercing beaks

In flesh eating and carnivorous birds the beak is pointed, sharp edged, hooked and strongly built as in hawks, owls, kites, eagles, vultures etc. The beak is used in tearing the flesh with the help of their powerful mandibular muscles.

8. Fish gripping beaks

Birds feeding on fishes develop long, powerful and sharply pointed beaks to catch fishes and other aquatic small animals. In some cases the edges of the beaks are armed with teeth-like projections . Ex. Kingfishes, snake bird, cormorants etc.

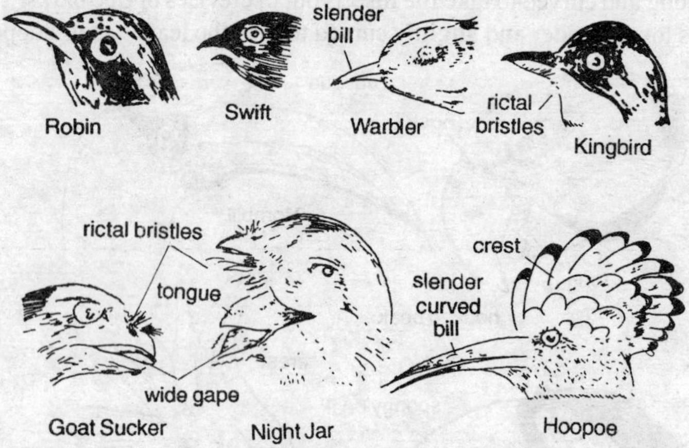

Fig. 19.3. Kinds of insectivorous beaks.

9. Flower probing beaks

Birds feeding upon the nector of flowers develop a long ponted beak depending upon the type of flower on which the birds feed. Ex. Humming birds.

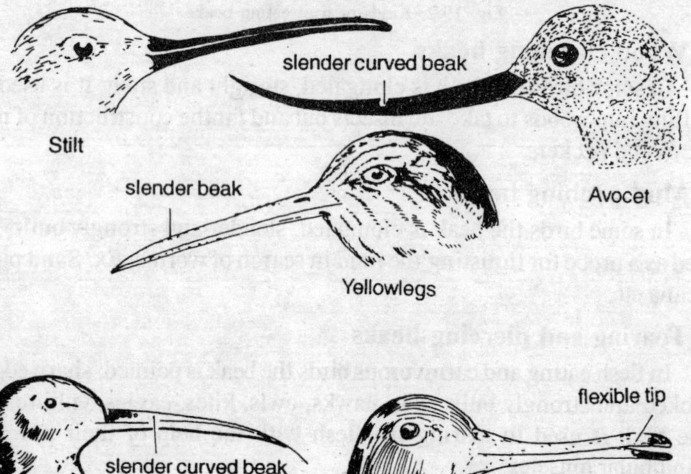

Fig. 19.4. Kinds of mud probing beaks.

10. Spatulate beaks

In birds like spoon-bill a flattened beak, that terminates in a broad spoon like expansion, is developed for dabbling in water and mud in search of insects, work, molluscs etc.

FEEDING ADAPTATIONS

Having completely lost the teeth the birds must depend largely on internal process to break up the food, the ingenciety of persistance with which birds seek and collect food must be a main factor in their success. Some of the most obvious differences among birds are seen in the numerous bill adaptations for feeding.

The flattened bills of Ducks are equipped with strainner (lamellae) for shifting out submerged food items. Hawks and owls possess powerful hooked beaks operated by well developed mandibular muscles for tearing up prey. Semiaquatic shore birds have long bills, extermely long in some for probing into mud and water, the probing bill of the woodcock in addition has a flexible tip on the upper mandible to faciliate siezing an earthworm deep underground. Wood peckers have strong chisel like beaks for piercing into wood for obtaining food or for nest construction, and have thickened shock absorbent bones of the head and strong neck muscles to make such pounding feasible. The familiar repai like probing bill of humming birds, amazingly modified to serve special needs in many tropical forms. The warblers have small wheras short stout bills for crushing seed are characteristic of seed eaters.

Like some fantastic plumage decorations, certain curiously constructed bills have probably evolved beyond the stage of any readily fare seen usefulness as in the enormous spongy beak of the toucans and the crested casque of the horn bills. On the other hand dextrally curved beak of the wrybill plover (*Anarhychus*) of New Zealand is said to function admirably for extracting prey for beneath the edges of stones which the bird circles in a clockwise direction. The peculiar bills of black-skimmer serves the unique skimming purpose implied in the name. The bills of extinct *Huias* of New Zealand were even more remarkable, that of the male was straight and sharp for tuneling into dead wood for grubs, that of the female was stender and decurved for extracting prey where the male could not reach. Then they shared the spoils showing in an usual example of co-operative hunting.

Feeding Mechanism

Some birds watch for the appearance of their prey and then descend upon it.

The lesser flamingo of Kenya skims the surface of the water with a semicircular motion of its inverted bill, and with the fine almost hair like serrations along the edges of the bill takes out the blue green algae and diatoms on which it subsits almost exclusively.

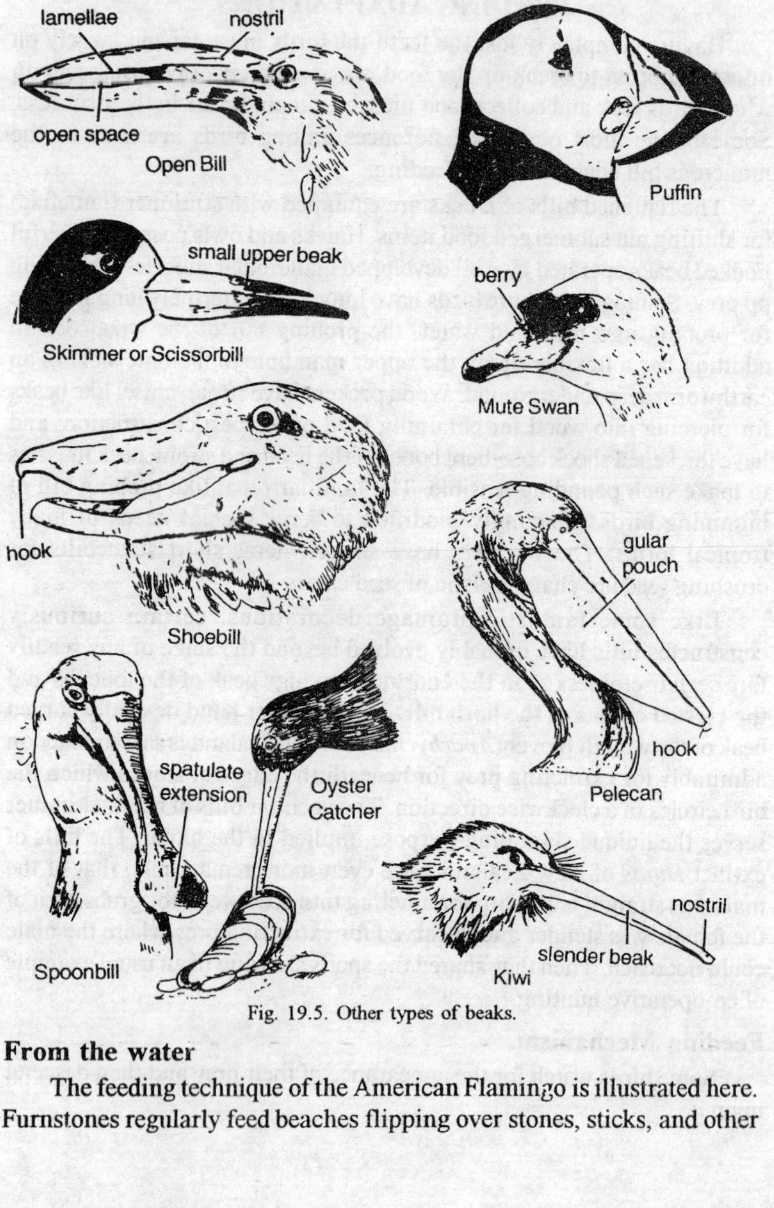

Fig. 19.5. Other types of beaks.

From the water

The feeding technique of the American Flamingo is illustrated here. Furnstones regularly feed beaches flipping over stones, sticks, and other objects by means of their strong, some what wedged shaped and stout beaks.

Darwins finches hold bills twig or thorn in the bill and thrust it into carniries of bark or into the holes it has itself exacavated when an insect emerges, the bird drops its probe and seize the insect.

CLAWS OF BIRDS

Like the birds beaks, their feet run through a variety of shapes and functions enabling them to live in different environment. Wood peckers have two, of the four bird toes, pointing back wards and with the help of very stiff feathers they serve to give the bird the extra perching it needs as it climbs vertical tree trunks.

Archaeopteryx had claws on all its fingers. Out in recent birds they occur on the first two only, being functionless in the adult or modern birds fingers much reduced in size and parts and almost none have claws. The ostrich has little or no claw on the outer toe. The manus had strong hooked claws at the tip of the fingers.

Differences between Avian claws and Human feet

The entire limb which supports the body and which has a foot at its extremity is called the "leg" by most people. The bony frame work of the

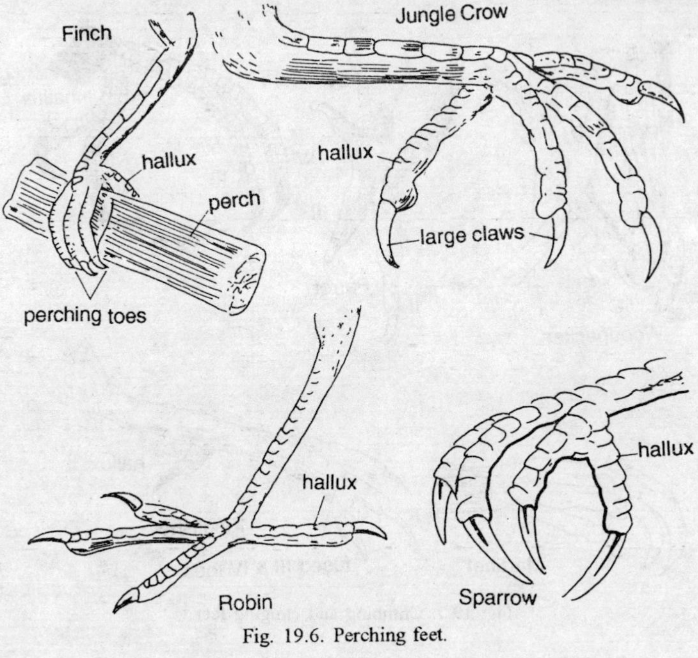

Fig. 19.6. Perching feet.

hind limb is a metatarsals and phalanges. The hind limb of a bird differs in several respects from the condition found in man. The ankle joint is not located between the two bones as in man, but lies between the two rows of foot bones. (Tarsal bones). During embryological development, some of the trasal bones fuse (grow together) with the distal end of the tibia. Hence the major supporting bone for the birds leg is not the tibia as in man but is more probably called the tibio tarsal a word combining the names of the fused bones. The fibula is second bone of the leg is poorly developed in birds, usually it does not fuse with tubia. The tarsometatarsus, which is roughly comparable to our instep is formed by the fusion of other tarsal bones with 3 meta-tarsal bones. A single independent metatarsal bone is present and serves as a support for first toe. The foot of the bird in broad sesne is composed of the tarsometatarsus and the phalanges (the bones of the toes).

Claws and Adaptations

Feet of birds are modified to serve specific needs. Birds feet are usually arranged with a back pointing first toe or hallux and forward pointing second to fourth toes. the swift running ostrich has lost all but its third and fourth toes; most other big flightless birds and some water birds have lost the first.

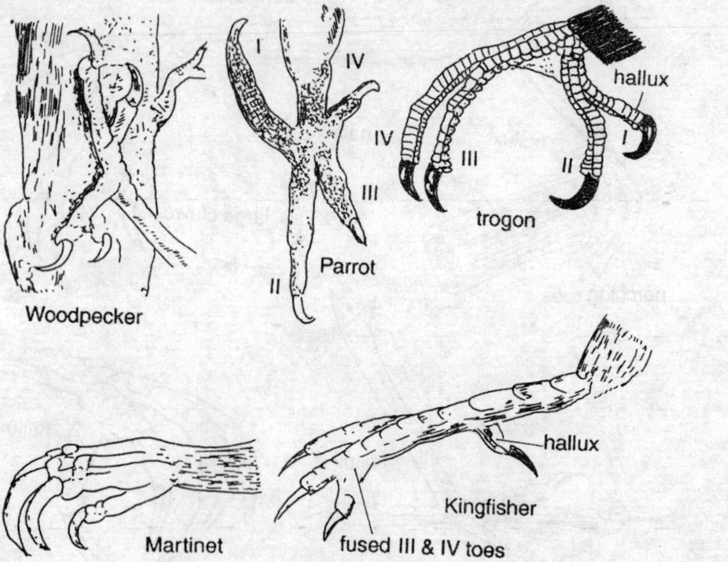

Fig. 19.7. Climbing and clinging feet.

Most swimming birds have feet modified as propellers or steering organs. The pelicans, cormorants and boobies have a "totipalmate" food in which the first toe is brought round to the side and a web embraces all four. A more normal "Palmate" footed duck has just the free front toe webbed.

Several orders of birds have the claws of their toe modified as preening combs e.g. are boobies, herons, bitterns and night jars. More terrestrial species like the skylark often has long hind claws to help stemces on flat ground. Nearly all the birds of prey bill either by striking or clutching with heavy sharp claws (Taloas). But the old and new world vultures never strike and have weaker feet.

Webbed or loabed for swimming in aquatic birds, long toed for walking over aquatic vegetation in some marsh birds, strongly taloned for seizing prey in predatory birds. Large hind claw of the lenks and certain frigates may have some functional significance. The pectinate claw (comb) on the middle toe are used for scratching the feathers of the head. Birds that walk over soft ground or an aquatic vegetation show

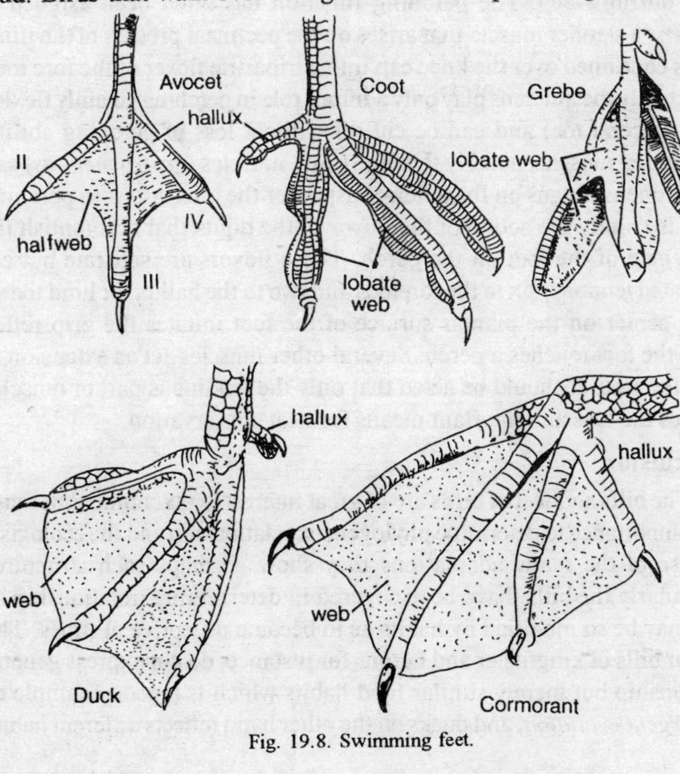

Fig. 19.8. Swimming feet.

similar adaptation to those of the herons. Passerine birds such as larks, pipits, and long spurs, which spent much time on the ground have elongated, claws, that of the hallux being specially long. Some of the concals (entropus) which live in swamps have also long clawed. The extremes in length of toe and claw are found in jacanas which spend much of the time walking over floating vegetations. In sand-grouse of the genus *Syrihaper* the feet show modifications for walking on sand. Both the toes and claws are short and broad. Nearly all the truegrouse show adaptations for walking on snow. In winter the claws are long and wide and a row of flat scales extend from each side of the toes. The claws and presumably the scales are moulted in the spring and grow tall again.

Perching Mechanism

Mainly the calf muscles are responsible to bring about flexion of the toes during perching and form an elaborate system of tendons attached to phalanges. These tendons act as a single unit and there is an interlocking arrangement by which the flexation is maintained by the weight of the body during sleep. The perching function has often been asigned to *ambiens* a slender muscle that arises on the pectineal process of the ilium and is continued over the knee cap into a tripartite flexer of the fore toes, but actually the ambiens play only a minor role in perching (mainly flexion of the second toe) and can be cut off without loss of perching ability. Several leg muscles notably the big "Calf" muscles (*gastrocnemius*) and the peroneus longus on the anterior aspect of the crus, assist in perching but it is chiefly the action of the *flexor* of the digits that accomplish the flight grip of the foot on the perch. These flexors are separate but co-ordinated tendons, six to the foretoes and two to the hallux, or hind toe. A nerve center on the plantar surface of the foot initates the grip reflex when the foot touches a perch. Several other muscles act as extension to open the toes. It should be acted that only the tendinous part of muscles reaches the toes an important means for heat conservation.

Conclusion

The bills and feet of birds are of great interest to taxonomists because their importance to determine phylogenetic relationship and the ecologists because of the many adaptations they show. Perhaps such structures particularly the bills, have been covered in determining relationship as they may be so modified by habits as to become deceptive at times. The similar bills of kingfisher and herons for instance, do not express genetic relationship but merely similar food habits which is a good example of *convergent evolution*, and ducks on the other hand reflects different habits and tend to show genetic relationship, a case of *divergent evolution*. Thus these structures play a diverse role in food gathering, nest building, defence and aggressive flighting in addition to the more precise function of merely eating and walking. The many fascinating adaptations illustrated greatly by bills and feet of birds are mere interest to beginning studens but there are largely a matter of observations in field and laboratory and need only to briefly summarised here.

20

The Origin of Birds

Of all the animals the birds are the best known and most easily recognised. The study of comparative anatomy and of the fossil evidences of ancient life have convienced zoologist that birds are unquestionably descended from repitles. The bird skeleton shows so many reptilian features that *Huxley* said that, 'Birds are glorified reptiles' and many zoologist have preferred to combine birds and reptiles in a single class the Sauropsida which means reptile like animals. It is also well known that the evolutionary potentialities of archosaurs expressed themselves not only in variety of adaptations to terrestrial and amphibian life but also in two experiments in flight represented by pterosaurs and ancestral birds.

The birds are far more reptilian in their organisation than the mammals. The evolution of feathers presumably from scales of reptilians type confered obvious advantages for the avian flight. The *Archaeopteryx* and *Archaeornis* thought bridge the reptiles and birds, but still the early fossil remains of birds are scare. Therefore, there is a still demand by the palaeontologists to have a transitional fossil to record exactly what evolution of birds actually means. Thus in few of sufficient transitional type between reptiles and aves, discussion of birds origin has taken the form of theories as to how flight got started among the birds. Several theories exist to account for this. Thus the origin of birds in current times in the absence of exact fossil link is generally dealt on the anatomical peculiarities of the *Archaeopteryx* and the theories of hypothetical proaves

Affinities of Birds with Archosauria

1. The birds are far more reptilian in their organisation than mammals. The structure of dinosaurs and of pterosaurs, and the birds shows many similarities which point the common origin of all three groups from the primitive thecodont stock.
2. The establishment of bipedalism (walking on two legs) among early thecodont left the fore limbs free for conversion into wings, by the ancestors of pterosaurs and the birds. But as the contrasting wing

structure in the two groups clearly shows that they must have arisen independently and evolved along the different line from the very fast.

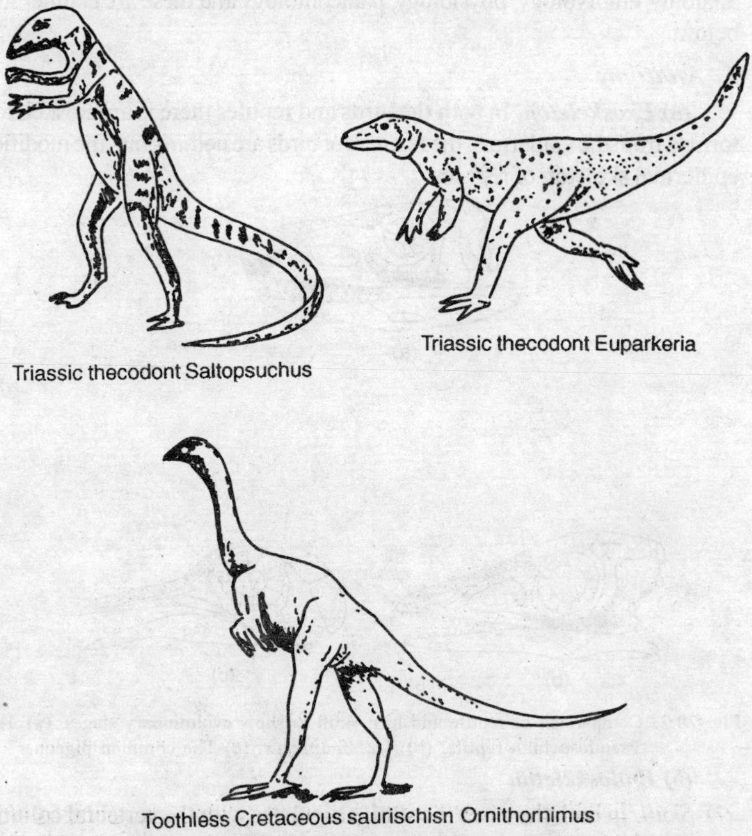

Fig. 20.1. Restoration of probable fore-runners of birds.

3. Resemblances such as the pneumatisation of the bones, the tendency to develop toothless beak which was present among some dinosaurs, the enlargement of the sternum and the form of brain are doubtless showing parallel evolution.
4. Birds are like the ornithischians dinosaurs in the structure of their pelvic but are more like sauroschain dinosaurs in their structure of both arms and legs and this fact misled early writers who believed that birds were descendent from the bird-like dinosaurs. But on the contrary the above dinosaurs lack a clavicle and they differed widely from birds in other skeletal peculiarities.

Affinities of Birds with Living Reptiles

Birds show many resemblances to the recent existing reptiles in the anatomy, embryology, physiology, palaeontology and these are enumerated below:

1. Anatomy

(a) Exoskeleton. In both the birds and reptiles there is an exoskeleton formed from the epidermis, the feathers of birds are nothing but, the modified epidermal skeleton of reptiles.

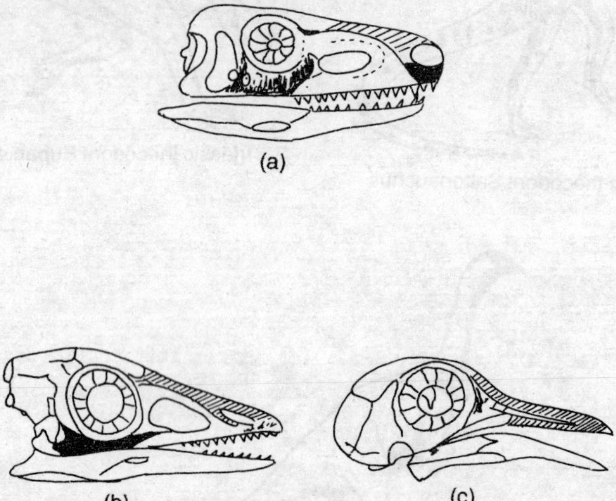

Fig. 20.2. Comparison of reptile and bird skull to show evolutionary stages. (a) The Pseudosuchian reptile. (b) *Archaeopteryx*. (c) The common pigeon.

(b) Endoskeleton

(*i*) *Skull.* In both the cases the skull articulates with the vertebral column by means of single occipital condyle. The diapsid condition of skull is formed in both reptiles and birds.

(ii) The lower jaw is similar in both the groups.

(iii) The suspensorium is biquadrate and is autostylic.

(iv) The epiphyses are absent in both the cases.

(v) The ribs, as in Rhyncocephalia, have uncinate processes.

(c) Visceral Organs

(i) Respiration in both is pulmonary and there is no gills in the adult. The air sacs which formed in the birds are also present in the lungs of *Chameleon*.

(ii) The heart is similar to crocodiles.

(iii) The red blood corpuscles are oval and nucleated.
(iv) There are 12 pairs of cranial nerves in both the groups.
(v) The eye is surrounded by sclerotic bones as formed in some dinosaurs.

2. *Embryology*

(i) Birds are oviparous as reptiles are, with certain exceptions.
(ii) The ova are very large with much yolk usually with calcareous shells.
(iii) Cleavage is meroblastic.
(iv) In early development amnion and allatois are very similar in both the groups.

3. *Physiology*

Experiments on the physiology made in the recent years show a close resemblance between birds and reptiles. The precipitation method of blood test shows a degree of genetic relationship between the two groups. They further indicate that the birds are more near to the turtle-crocodile series than the lizard-snake series.

Archaeopteryx as Palaeontological Evidence

The oldest fossil bird *Archaeopteryx* which is more reptilian than birds furnishes a good example of reptilian ancestory of the birds. The earliest known birds appear rather late on the scene than the pterosaurs towards the end of the Jurassic. Only two specimens and one isolated feather are known to the palaeontologist. These were discovered in State as in a quarry located near Pappenhiem, Bavaria. The first skeleton was found in 1861 and was named as *Archaeopteryx lithographica*. The second skeleton (now in Berlin) was found in 1877 and has been called *Archaeornis siemensi*. The skeleton of British museum of Nautral History has been restudied by *De Beer* (1954). By the use of new techniques such as the examination under ultraviolet rays that these two skeletons were specific.

The skeleton of *Archaeopteryx* is more reptilian than the avians and infact *Lowey* (1944) denied that *Archaeopteryx* was bird. The despite of *Lowey* assertions, however, the avian affinites of *Archaeopteryx* are now generally held.

These two (*Archaeopteryx* and *Archaeornis*) were the Jurassic birds. The Cretaceous birds found as fossil are *Ichthyornis* and the other *Hesperornis* both were marine forms. The *Hesperornis* was found in the marine Niobarara Cretaceous chalk pads of Kansas. This was described by *Marsh* (1880) and these birds have been spoken as the birds of Cretaceous or the toothless Cretaceous birds. It is generally agreed now that both birds and dinosaurs are descendant from primitive unspecialized thecodont reptiles belonging to sub-order Pseudosauchia. These were small carnivore possessing many teeth in sockets in the jaws. Their hind-limbs were larger than their front limbs and in some forms the 5th hind toe was reduced in length.

THEORIES OF FLIGHT

1. Nopcas theory or Cursorial origin of flight

Cursorial means fast running forms. The theory of the cursorial origin of flight was advanced by *Nopcas,* a Hungarian palaeontologist. *Nopcas* considers that there is a very fundamental distinction between flight based on membranous plumes like those in bats and pterodactylus. The plumage is made up of feathers. For the former involves marked adaptation of the hind limbs, whereas the latter involves only the fore limbs and leaves the hind limb unchanged. The hind limbs of birds are essentially homologous with those of cursorial dinosaurs. It is, therefore, argued that the origin of flight occured while running efficiency was at its height. The conclusion is that the first birds arose from some long tailed reptiles.

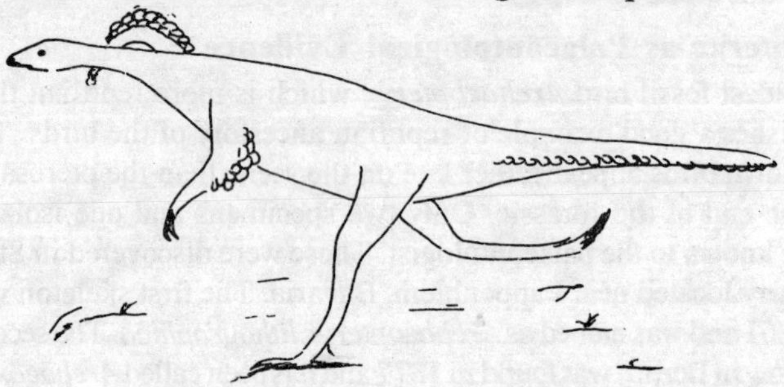

Fig. 20.3. Restoration of a hypothetical pro-aves.

A hypothetical *pro-aves* that spread over the earth on its strong hind legs, stretched out its fore limbs for the sake of maintaining balance and probably flapped. To aid the speed of running, these flapping fore limbs or pro-wings developed more surface, partly by flattening out and partly by the backward growth of scales of the posterior margin. Similar large scales are supposed to have developed laterally on the tail. Thus evolution of specialized flight scales into feathers is thought to have been a more matter of a continued increase in size and numbers, accompained by regional specialization. According to this theory, 'A feather is regarded as more or less a specialized scales.' The gradual modification of the remaining body scales into feathers would be the logical sequence of events to the long list of flight adaptations would appear as correlated variation.

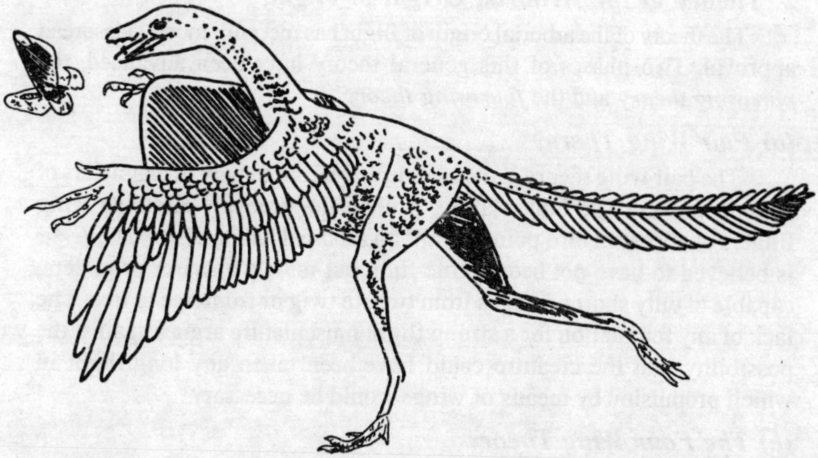

Fig. 20.4. Restoration of *Archaeopteryx*.

Fig. 20.5. A—The arboreal pro-aves, B—Arboreal *Tetrapteryx*.

The first step in flying would be prolonged leaps aided by flapping pro-wings then short souring flight, would be made followed by larger flight, accomplished by energetic flapping of the wings alternating with periods of soaring. While rather accepted some ways the theory of the cursorial origin of flight has not gained general acceptance.

2. Theory of the Arborial Origin of Flight

The theory of the arborial origin of flight has met with more widespread approval. Two phases of this general theory have been advanced, the *pair wing theory* and the *four wing theory.*

(a) Pair Wing Theroy

The pair wing theory is derived directly from a study of characters of *Archaeopteryx*. The long clawed, prehensile, probably climbing wing fingers of ancestral bird point toward an arboreal habitat. *Archaeopteryx* is believed to have not been a true flier, but merely a soarer or gliders, capable of only short passages from twig to twig or from tree to tree. The lack of any foundation for a strong flight musculature argues against the possibility that the creature could have been taken any long flight in which propulsion by means of wings would be necessary.

(b) The Four Wing Theory

The four wing theory is of *Beebe*, is the most recent theory, dealing with the origin of flight. This author made the remarkable discovery that vestigial flight feathers occur on the thigh of a number of species of modern birds. Traces of similar feathers were found on the thigh of *Archaeopteryx*. These discoveries led to the conclusion that the first flier had wings on both arms and legs and used both sets in gliding from trees to the ground and from trees to trees. Latter the origin of the legs wings

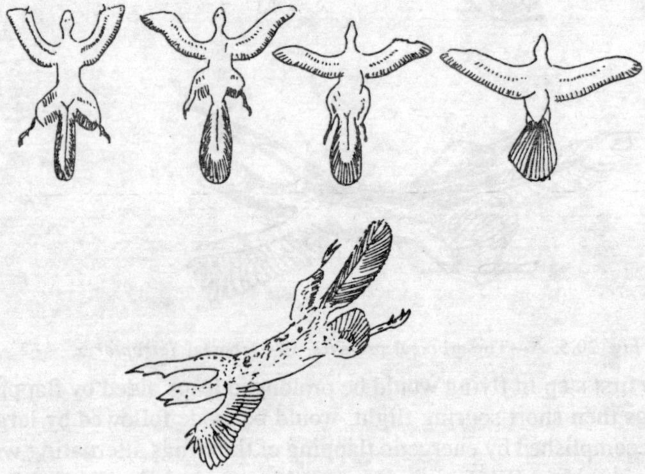

Fig. 20.6. Four wing theory of origin of flight.

degenerated as the tail feathers took up the duty of acting as a posterior plane and effectiveness as motor organs.

3. Gregory's Compromise Theory

Gregory's compromise theory of the origin of flight is perhaps more nearly acceptable than any of those hitherto given, and is here presented in his own words.

Fig. 20.7. Reconstraction of arboreal *Archaeopteryx* compared with modern birds.

The pro-aves were surely quick runners, both on the ground and on the trees, but it is not clear whether the upright position was first attained upon the ground and on the trees. They very early acquired the habits of searching upright on the.branches as shown by the consolidated in step bones, grasping first digt and strong claws of *Archaeopteryx*. Their slender arms ended in three long fingers provided with large claws which were at first doubtless used in climbing. These active pro-aves contrasted widely in habit with their sluggish remote reptilian fore bears. In pursuit of their prey they jumped lightly from branch to branch and finally from tree to tree, partly sustained by the folds of skin on their arms and legs and latter by the long scaly feathers of the pectoral and pelvic wings and tail. That they held the arms and legs perfectly still through out the gliding leap appears doubtful, far all recent animals that do have never attained true flight. It can not avoid the impression that a vigrous downward, flap of the arms even before they became efficient wings would assist in the take off for the leap and that another flap just before landing would check the speed and assist in true landing.

Gregory in 1916 stated that the birds had a two fold or dual origin, some from cursorial (fast running) forms and other from arboreal forms. It has conclusively proved that the birds have descendent from the reptiles.

4. Diving Origin of Flight

According to this view the pro-aves used the fore limbs together with their membranous and elongated scales or plumes to aid in diving. The value of such accessories obvious, the dive being more definitely directed the descent mode flatter so as to carry the diver farther out from shore and the force of the plumage eased up sufficiently to avoid shock. If the wings were flapped more or less, a longer glide out over the water could be made and possible circling movements could be taken over the water while searching for fish. It would appear, therefore, that the prowings as placed in diving would serve as useful function as in running or leaping. We would then have to support that some of the arctics diving birds such as the Penguins underwent a specialization of the primitive wings using them for underwater flying.

The strong flying sea birds would then be derived from ancestral diving types that had gradually perfected their flight, while land birds of all sorts would be derivatives of the sea birds. There are infact many evidences that the sea is the ancestral home of the birds and that they have invaded the land in comparatively recent time. *Archaeopteryx* of course seems to indicate against the diving origin of flight for it is assumed to be a climbing arboreal bird. But might not climbing the equally appropriate as an aid in sealing puffs after diving and swimming in the water. On the whole the existence of *Archaeopteryx* is no more a barrier to acceptance of the diving than of the arborial origin of flight when other considerations appear to make the former more probable than the latter.

Conclusion

We may conclude the origin of birds by stating that far beyond doubt that the birds originated in the Triassic times from a bipedal cursorial reptiles which had an elongated body axis and a lizard like form. The wings were developed at the sacrifice of the fore limbs and the scales were transformed into feathers, simply by a process of splitting up. The strongest evidence is provided by the fossil bird *Archaeornis*.

21

Breeding and Nesting

In reproducing their kind, birds frequently show great ingenuity, whether it's the blackbird nesting in the garden shed, the stork on the chimney pot or the falcon on a sky-scraper. Many have adapted well to a world now dominated by man, but millions of years before *Homo sapiens* first walked the earth birds had already learnt to build nests and rear young in virtually every natural niche on earth. Despite the rigours of having to protect fragile egg in frequently hostile environments, they nest in every continent, from arctic ice to waterless deserts where no man could endure. All lay eggs, but not all make nests. Some are born blind, naked and helpless, but other hatch 'ready to go'. Most are good parents, but others show surprising lack of care and a few parasitic species have learnt to hand the entire business of baby care over to foster parents.

The Largest Tree Nest in the World

Bald Eagle's - up to 2.9 m (9.5ft) wide and 6 m (20.0 ft) deep : If we take the popular meaning of nest — a structure made by a bird from natural materials such as twigs and grasses, and containing a cup or cavity in which eggs are laid or incubated — then the largest are undoubtedly made by some of the birds of prey which use the same site year after year. The very largest of these are generally made by the bald eagle *(Haliaeeius leucocephalus)* and the biggest known as 2.9 m (9.5 ft) wide — more than enough for the tallest man in the world to lay across with his arms outstretched — and an incredible 6 m (20 ft) deep. This monotrous fortress at St Petersburg, Florida was almost certainly added to by several pairs, the usual maximum depth for any — even the largest eagle eyrie — being about 3 m (10 ft). The next weighted more than 3,000 kg. (2.9 tons). Another bald eagle nest at Vermilion, Ohio reached 2.59 m (8.5 ft) across and 3.66 m (12ft) deep after 35 years. Eventually it crashed to the ground during a storm, killing the egglets inside. It had an estimated weight of 1,814 kg. (1.78 tons).

Most huge bird-of-prey nests occur in remote or inaccessible areas where the birds have nested undistributed for centuries. Those of the Australian white-beilled sea-eagle *(Haliaeetus leucogaster)* may be 2-3 m (7-10ft) across and 4 m(13 ft) deep and traditional nests of the osprey along the Red Sea coast are atleast 1 m (3.25 ft) across and 2 m (6.5 ft) high. The largest nests in Britain are those of the golden eagle *(Aquila thrysaetos)* : in 1954 British naturalist *Seton Gorden* saw one in Scotland which was 4.6 m (15ft) deep and had been in use for 45 years.

The World's Largest Nest-Mounds

Scrub fowls, big footed relatives or the pheasants *(Phasiantidae)*, mate for life and a pair inhabits a large territory. It is mostly the male, which works the mound. This begins with the digging of a hole up to 5 m (16 ft) in diameter and 1- 1.5 m (3 - 4.5 ft) deep. The mound may contain 3,00,000 kg (295 tons)of matter and the birds can move very large objects. A 1 kg (2.2 lb) scrub fowl was once seen to shift a rock of 6.9 kg (15.2 lb). During the winter this hole is filled with vegetation swept up from the ground over a radius of about 45 m (150 ft). When this is moistened by rain the whole is covered by a sandy soil 50 cm (20 in) or more thick. Sealed from the air, by solar heat. When fermentation heat is too much in the spring the birds need to open the mound frequently to let the heat escape and maintain the ideal incubation temperature of 34° C (93.2° F). In midsummer the suns's heat is most important, but this may be too much and then the thickness of soil has to be increased to prevent 'overcooking'. But by autumn the sun's heat is less and the fermentation heat is exhausted so the birds open and flatten the mound daily to let the sun's rays warm the eggs more easily. In this way the birds achieve precise temperature control over several months.

None of the 12 species megapode of Australia, New Guinea, Indonesia and Polynesia uses body heat to incubate eggs. The required heat comes from the decomposition of vegetable matter in the mounds, directly from the sun or even from volcanic upwellings. Where the mound temperature is naturally fairly constant some species leave the eggs to hatch on their own, but most megapodes employ great skill in maintaining a remarkably constant incubation temperature so that the eggs are not spoiled through chilling or over- heating. The precise way in which they do this is uncertain, but one suggestion is that they gauge the incubation temperature with ultra-sensitive tongues.

In some areas scrub fowls are regarded as rain-prophets as they start to scratch their mounds together at the approach of rain in spring. But they will also begin work if rain falls in the autumn and will attend to mounds through practically the whole year, raking them over for essential temperature regulation and aeration.

Direct solar heat may be more important for the well-studied mallee fowl (*Leipoa ocellata*) of southern Australia as it breeds in dry places where, in the absence of rain, rotting of mound material is very slow. But for those species which live in dense, susnless jungles heat from decomposition is crucial. Yet some moisture is important in everymound to prevent the egg shells cracking.

The number of eggs laid varies from 5 to 35, the great variation being not only between females but also in the same individual from year to year. The interval between the laying of eggs depends on the nutritional state of the female but is often several days. They may be laid over several months but each begins the nine-week incubation immediately so that some chicks may leave the mound before the last eggs are laid.

The Largest Social Nest Sociable Weaver's-up to 300 chambers: The weaver family is well-known is well-known for its 'hanging sock' nests with tubular entrances as long as 70 cm (27.5 in) and as much as 10 cm (4 in) wide. One of the most remarkable is that of thesociable weaver (*Philetairus socius*), of south-west Africa which first constructs a roof of coarse straws in a tree or on a telephone pjole and then makes as many as 300 nest chambers below. The ensuing mass can dominate an entire large tree and attract other species. The tiny pygmy falcon (*Polihierax semitorquatus*) is one guest but, although it feeds on small birds, it does not molest its hosts.

Probably the most remarkable of the social-breeding parrots is the South American monk parakeet (*Myiopsitta monachus*) which builds huge colonial twig nests in thetops of trees. Within the main structure each pair has its own nest chamber. In Birds of La Plata (1920), *Hudson* noted that such a clump can 'weigh a quarter of a ton and contain enough material to fill a large cart'.

The largest roofed nest hammerhead stork's -up to 2 m (6 ft) wide and 2m (6.5 ft) deep : More popularly known as the hammerkop, the African hammer-head stork (*Scopus umbretta*) defies accurate classification and is not closely related to the sotrks. It prefers to build its huge nest of twigs, grass and mud about 12-15 m (40-50 ft) up in a tree. an though the nests are usually 1-2 m (3.25-6.5 ft) acros, often of similar depth and take some two months to make, this industrious bird may make several in one season. No one knows why it goes to such great lengths to raise its young.

First a platform is constructed of sticks cemented with mud. Then the wall are raised and roofed over with a metre of thatch which may be decorated with feathers, bones, snakeskins and other debris. The whole thing may contain over 10,000 sticks and easily support the weight of a man. The internal chamber, which is about 30-50 cm (12-20 in) in diameter and height, is verycarefully shaped and large enough to house both parents and yound. The entrance tunnel is 13-18 cm (5-7 in) wide and 40-60 cm (16-24 in) long and heavily plastered with mud to get in. Where trees are scare the nest may be built on a cliff face or the gound, but man-made structures such as dams are also used.

The three to six young, fed by both parents, spend about seven weeks in the nest before fledging. When both birds feed the yound they may leave them for long periods, and presumably the thick nest walls then provide good protection. Even after fledging the yound remain near the nest for a month or so roosting in it at night. Many other species are also attracted to the massive nests, whether occupied or not. During construction some nests have been take over by Verreaux's eagle (*Aquadla verreauxii*), and grey Kestrels (*Falce ardostaceus*) and barn owls (*Tyto alba)* often evict the rightful owners from the finished nests. Many smaller birds such as weavers, mynahs and pigeons attach their own nests to the main structure. Other hole-nesting species such as the Egyptian goose *(Alopochen aegyptiacus)* and African pygmy goose *(Nettapus auritus)* soon take over the vacated nests, so overall the hammerkop provides important nest-sites for many birds which might otherwise have to fore-sake the habitat.

Small Builders of Big Nests

The Most extraordinary large nests for small birds are made by oven-birds *(Furnariidae)* (10-26 cm / 4-10 in long). That of the rufous-breasted spinetail *(Synallaxis erythrothorax)* is made from thorns, with tunnels to the nest chamber. It is roughly oblong or retort-shaped and as much as 75 cm (30 in) long by 50 cm (20 in) wide. It has a large platform on the side which leads via a tunnel through a tangle of thorns to a thatched nest-chamber lined with downey leaves. That of the rufous-throated thornbird *(Placellodomus rufifrons)* has further chambers added in subsequent seasons and gives the impression of a large colonial nest. It is unlikely that spare chambers are used other than by non-breeding members of a previous brood. Also voluminous is the thornyn nest of the 21 cm (8.3 in) firewood gatherer *(Anumbius annumbi)* of Argentina. It measures about 70 cm × 30 cm (28 in × 12 in) and is made of big sticks with a crooked passage at the top leading down to the nest chamber. But the largest in this varied family is probably the huge, thorny nest of the white-throated cachalote *(Pseudoseisura gutturalis),* being up to 1.5 m (5 ft) in diameter. The enclosed structure has a cavity big enough for an eagle or vulture and strong enough for a man to stand on without damaging it. The brown cachalote *(Pseudoseisura lophotes)* of Paraguay, Uruguay and Argentina makes a nest the size of a barrel. Famous naturalist *W.H Hudson* noted that he could stand on the nests of some of these species and make no impression. Although these huge nests are conspicuous feature of open South American countyside little is known about the breeding because of the impenetrability of the structures.

The longest nest burrow the Rhinoceros Auklet's 8 m(26 ft): Why the 36 cm (14 in) Rhinoceros auklet *(Cerorhinca monocerata)* of the North Pacific should excavate such a long burrow is a mystery. This misnamed bird is actually a puffin and its range is from the coast of Kamchatka and the kuril islands and from the west coast of Alaska south to Washington. In winter it is found in Japan and lower California. Its single egg hatches after about 40 days and the chick fledges after 40-50 days, having been fed on fish.

The Strangest Nest-Sites

There are many fascinating nest-sites across the world, but the strangest must include that of the violaceous trogon *(Trogon violaceous)* of Central and South America. This bird takes over a wasp's nest, eats the adult wasps, then digs out the comb to make a nest cavity.

The water thik-knee or dikkop *(Burhinus vermiculatus)* of Africa is equally unusual in frequently laying its eggs on the dried droppings of large mammals such as the hippopotamus.

The Highest Tree Nest

The Marbled Murrelet'ss 45 m (148 ft) : Kinglets (*Regulus*) and Lewis's woodpecker (*Melanerpes lewis*) nest at 30 m (100 ft) up in trees but the highest nest found so far is that of the marbled murrelet *(Brachyramphus marmoratus)* at 45 m (148 ft). A very strange situation indeed for a 24-25 cm (9.5-10in) auklet. The single egg is laid in a small cup, made mainly of guano, on a pad of moss in large coniferous trees in remote areas upto 10 km (6 miles) inland. Only five nests of this common seabird have ever been reported. Whereas other auks breed in colonies, often huge, only the marbled and Kittlitz's murrulet *(Brachyramphus brevirostris)* nest in solitary pairs. No one had ever seen the nest of this bird until 7 August 1974 when one was found in a Douglas fir in Big Basin State Park, Santa Cruz Country, California. Kittlitz's murrulet is also exceptional for an auk in that it lays its egg on bare ground above the tree-line and far from the sea.

North American sequoias are the tallest trees in the world and may exceed 90 m (300 ft), but their tops seem to be avoided by birds and there are no records of any birds nesting at such a height.

The Smallest Cup Nest in the World

The bee hummingbird's-thimble- sized : There is no doubt that the tiniest cup-shaped nests are made by hummingbirds. These are mostly fixed to twigs by cobwebs and built by the female alone. The narrowest is the thimble-sized cup of the world's smallest bird — the bee humming bird *(Calypte helenae)* of Cuba and the isle of Pines. But this is slightly deeper than that of the vervain hummingbird *(Mellisuga minima)* of Jamaica and Hispaniola, whose nest of lichen, silk and cotton is only about the size of half a walnut shell.

The female humming bird alone incubates the eggs and cares for the young. Except in three or four species, male humming birds do not even know the whereabouts of the nests. Some, such as the hermit humming birds, build hanging nests, attached by cobwebs to the undersides of large leaves, such as palms and ferns, so that the leaf forms the inner wall and they are sheltered from tropical downpours. Others, such as lancebills, metal-tails and hillstars, use cobwebs and glue to suspend their nests from the ceilings of caves or rocky overhangs. The deepest humming bird nest was the only one known of the blue-fronted lancebill *(Doryfera johannae)*, which was pendent structure attached by cobwebs to a rocky overhang near the bottom of a 75 m (250 ft) shaft.

Many birds make no nest at all and lay their eggs on bare rock or earth, or even in a tree cavity. Some scrape together a few woodchips or merely make a depression in vegetation or shingle. The 25 cm in) fairy tern *(Sterna nereis)* of Australasia lays its single egg on the rare branch or leaf of a tree, or on coral boulders and cliffs. In the absence of a nest to contain it, the young has strong claws which enable it to cling tightly and even rang upside down.

Palm swifts also make tiny nests, one of the smallest being that of the 10.5 cm (4 in) pygmy palm swift *(Tachornis furcula)* of Colombia and Venezuela. The Asian palm swift *(Cypsiurus batasiensis)* and African palm swift *(Cypsiurus parvus)* build any open nests on the insides of palm leaves (they are restricted to where the fan palm grows). In Africa the minute eggs are glued to the nest with saliva. The nest is made from feathers and fibres and hasa small lower rim for the bird to perch on while it incubates vertically.

The Most Valuable Nests in the World

The cave swiftlets' : Cave swiftlets (*Collocalia*) are best known for providing the Chinese with their esteemed birds' nest soup. While all swifts and swiftlets use saliva to glue their nests together, species of *Collocalia* use saliva as the principle ingredient, though some incorporate feathers, leaves and other materials in varying proportions. Those most prized for soup making are made of pure saliva, and include the little nests of the grey-rumped swiftlet *(Collocalia francica),* which do not require extensive cleaning. The crop of such nests may be both valuable and extensive. In one year alone over 3.5 million nests were exported from Borneo to China. The nest sell for about £ 3 each and it takes two to make a single bowl of soup selling for up to £ 10 a bowl. The nests have no nutritional value and tasteless, but are highly regarded as an aphrodisiac.

The colonies in vast Asian caves my contain several hundred thousand individuals whose nests of dried saliva are often stuck to the roofs and walls of caverns in total darkness up to 400 m (1300 ft) or more from the entrance. Collection of them can be very hazardous, involving ropeways and ladders up to 100 m (330 ft) high. But the swiftlets have no trouble in negotiating the bird-filled blackness because they have developed the faculty of echolocation to a higher degree than any other bird species.

The well-studied edible-nest swiftlet *(Collocalia inexpectata)* has two peaks of maximal breeding activity : October- December and February-April, andthe nests take about 30-40 days to build. Even without the cropping by humans, breeding success is poor as they can fall prey to cave crickets which eat the eggs and sometimes the young. Many other nestings fall and are eaten by snakes, while those in houses may be eaten by rats and shrews.

The vast colonies also provide another valuable resource — the droppings (guano) which accumulate on the cave floor and are mined for use as fertilizer. But neither this nor taking the nests endangers te swiftlets' population. Rights to the caves are jealously guarded and nest removal has been carefully controlled for many centuries, even long before the first white man set foot in the area.

The Largest Egg of Any Living Bird

The Ostrich's - up to 17.8 cm (7 in) × 14 c m (5.5 in): The largest egg of any living bird and the biggest single cell in the animal world today is that of the ostrich (*Struthio camelus*), the world's largest living bird. There is considerable variation among the races but the North African birds generally lay the largest eggs, up to 17.8 cm × 14 cm (7 in × 5.5 in).

As with all species, egg-size is linked to individual female size, heavier individuals producing larger eggs, and is genetically controlled and inherited. Thus exceptional eggs do occur throughout Africa. One from Lake Jipe at the south-west corner of Tsavo (West) National Park in Kenya was 17.1 cm (6.7 in) long, 13.6 cm (5.35 in) across, 49.1 cm (19.3 in) round the long circumference and 43.7 cm (17.2 in) round the short circumference. It weighed 1,974 gm (4.35 lb) fresh weight. This was the largest of 600 eggs weighed and measured in East Africa and considerably larger than any of 100 measured on an ostrich farm at Oudtshoorn in South Africa. Most North African ostrich eggs are about 15.5 cm (6.1 in) long by 13.5 cm (5.3 in) across and weigh about 1.65 kg (3.63 lb).

Ostrich eggs are very strong, with a shell some 1.6 mm (0.063 in) thick, resisting the attentions of most predators, even playful lions, but Egyptian vultures (*Neophron percnopterus*) have learnt to crak them with stones. One egg is equal in volume to some two dozen hens eggs and takes about 40 minutes to boil.

Despite its great size, the ostrich egg represents only about 1% of the female's body weight. Other species lay eggs which represent up to 25% of body weight. Smaller specimens are laid by females breeding for the first time, but the oldest birds sometimes revert to laying smaller eggs. Scarcity of food may also restrict the size of eggs and small eggs are less successful at producing strong fledglings.

Large as they are, ostrich eggs are much smaller then preserved eggs of the extinct *Aepyornis maximus.*

The Largest Egg Proportionate to the Layer

The little spotted Kiwi's- 25% of the female's weight : As a proportion of the female's weight, the largest egg is laid by the little spotted kiwi (*Apteryx owenii) of* New Zealand (now known on only three offshore islands). The average 35 cm (113.8 in) female has a body weight of about 1.2 kg (2.6 lb), about 20% heavier than the male, and her eggs weigh 275-370 g (9.7-13 oz), with an average of 310 g (10.9 oz) - about 25% of her body weight. Average dimensions are 108.5 × 71 mm (4.27 × 2.79 in). Eggs of the North Island brown kiwi (*Apteryx australis*) are larger at 330-520 g (11.6-18.3 oz), with an average weight of 440 g (15.5 oz) (125.5 × 78.5 mm 4.94 × 2.76 in), but represent just 18-22% of the female's body weight.

Eggshell thickness varies considerably between species and this significantly alters the internal volume. The shell of the larger emu (*Dromaius novaehollandiae*) egg (635 g/22.3 oz, 525 ml/1.05 pt) is about four times and the shell of the ostrich (*Struthio camelus*) egg (1,240 g'43.7 oz, 985 ml/1.97 pt) about eight times the weight of the kiwi eggshell. Total weights, therefore, exaggerate the size differences between the kiwi egg and the eggs of larger ratites, as the weight of the ostrich egg is nearly three times the weight of the kiwi egg, but its internal volume is only about 2 11, times as great. The average 53 g (1.86 oz) domestic chicken egg and a 435 g (15.3 oz) kiwi egg are both laid by birds weighing about 2.4 kg (5.29 lb). Not surprisingly, kiwi eggs take a long time to incubate 65-85 days, almost entirely by the male.

The little spotted is the smallest of the three species of flightless kiwi. The largest is the Stewart Island race of the brown kiwi (*A. a. lawryi*), which has females of at least 3.5 kg (7.7 lb) and is the only kiwi active by daylight. Kiwis are much smaller than the other ratites because they have no natural predators in New Zealand and therefore size is unimportant in defence. The high-protein diet also makes size is unimportant. As well as worms, this includes woodlice, millipedes, centipedes, slugs, snails, spiders and a wide range of insects, as well as seeds and berries - all highly nutritious.

Captive females have laid up to five eggs in a continuous series with an interval of about 33 days between each. In the wild, clutches are usually 1-2 eggs (rarely 3) and the egg is heavily yolked (61%) like those of other species whose chicks hatch active, open-eyed and fully-feathered.

Even though hen ostrich lays 12-15 of the largest eggs in the world, weighing in total 20-25 kg (44-55 lb), such clutches represent only 20-25% of body weight. Hummingbirds lay two eggs, each of which weighs about 13% of the female's body weight, but the proportion is even greater in the smaller, species.

In total contrast, the smallest of all clutchs as a proportion of body weight is that of the emperor penguin (*Aptenodytes forsteri*) whose single-egg clutch represents a mere 1.4% of the female.

The Smallest Egg in the World

The bee humming bird's - 6.35 mm (0.25 in) long: In terms of average weight or volume, the question of which hummingbird lays the world's smallest egg is unresolved as comparatively few eggs have been comprehensively and accurately measured, and among these there has been considerable variation. But in terms of length alone the egg of the world's smallest bird - the bee hummingbird (*Mellisuga helenae*) of Cuba and the Isle of Pines - is the shortest, varying between 6.35 mm (0.25 in) and 11.4 mm (0.45 in) long. Those of the vervain hummingbird (*Mellisuga minima*) of Jamaica and Conave Island, Hispaniola are generally no smaller than 10 mm (0.39 in) long and 0.365 g (0.0129 oz) in weight.

However, abnormally small hummingbird eggs have also been reported for several species. These 'sports' or 'runts' are fairly common throughout the bird world and are generally thought to have been laid prematurely. One exceptionally small example of the Costa's hummingbird (*Calypte costae*) of south-west USA and north-west Mexico measured only 7.36 mm (0.29 in) by 5.33 mm (0.21 in) and contained no yolk. In the same nest at Escondido was a second egg which was slightly incubated and measured 12.7 mm (0.5 in) by 8.4 mm (0.33 in) Amazingly, the finder managed to blow the tiny egg and it is now preserved in the Western Foundation of Vertebrate Zoology at Los Angeles, California. The bird which laid the two eggs was just 8.6 cm (3.4 in) long.

Hummingbirds generally lay two (sometimes one) elongated, white eggs which may be very small to us but represent about 13% of the female's bodyweight. Incubation takes 14-23 days and the nestling period is 18-38 days.

The Largest Clutch

Grey *Partridge's — average* 15-19 eggs : The largest single clutches are laid by those species with a high mortality and low average life expectancy. This is particularly marked among the ground-nesting partridges, pheasants and grouse, some species of which frequently lay over 20 eggs. The bobwhite quail (*Colinus virgininianus*) is often said to lay the most eggs, and indeed it produces between 7 and 28, but its average is just 8-15. This is considerably less than the wide-spread grey partridge (*Perdix perdix*), which usually averages at least 15- 16, and in Ostrobothia, Finland the average is 19. Within many species clutches become larger with increasing latitude and longer summer days as there is a shorter breeding season with less time for successive broods but more daylight in which to feed a large family. The largest grey partridge clutch to a single hen reported in Britain was 25 in Sussex in 1974. Twenty-four chicks hatched and all were still alive at six weeks. Although it is through that clutch-size in most species is linked to the physical condition and age of the hen, it now seems that the amount of food the parents can collect for the potential brood is more important. There is also evidence that clutch-size is hereditary and that some individual females are either more experienced or more capable than others in caring for large broods.

Hole-nesting birds often lay more eggs than closely related species in more open sites, presumably because the enclosed eggs and young are safer. For example, the blue tit *(Parus caeruleus)*, which has adapted so well to the widespread use of nest boxes, generally has the largest clutch of any nidicolous (having young birds that remain in the nest after hatching) species. An average of 11-12 eggs is normal for good habitats in Europe, but individuals may lay up to 19. In tits and many other species there is also some seasonal variation in clutch-size, the larger clutches generally being laid early to benefit from peak food abundance.

Many supposedly large clutches are due to 'dump-nesting' where two or more females have laid in the same nest. This is particularly common among game birds and other *Galliformes* as well as ducks, and the generally unmanageable nest is usually abandoned. This sometimes happens when females have no nest of their own or occasionally practise parasitism. Incubation is very occasionally attempted, though not when the egg pile reaches, intimidating proportions. For example, one North American redhead (*Aythya americana*) nest contained 87 eggs! Some redheads are entirely parasitic, making no attempt to build a nest and laying all their eggs in the nests of other duck species. Others are partially parasitic and lay eggs in the nests of other ducks, before settling down to raise their own family. Yet others never practise parasitism at all.

Some species, such as anis (*Crotophaga*) nest communally and share the incubation. This is quite distinct from dumpnesting. Up to 29 eggs have been found in one ani nest and the socially dominant pair does most of the breeding. The anis of the Americas are related to the parasitic cuckoos (*Cuculidae*).

Ostriches (*Struthionidase*) have a fascinating system of communal nesting. The cock defends a large territory and acquires a mate who lays her eggs in a scrape. A few days later other hens arrive and also lay in the nest. Surprisingly, the first, or major, hen gets off her nest to let them in, but only she and her mate will incubate - up to 40 eggs! But even an ostrich cannot cover so many eggs and some get pushed to the edge. This was once thought to be a way of regulating the temperature of the eggs, but investigation has shown that the outer eggs get overcooked in the sun.

The major hen is able to recognize her own eggs (perhaps by the pore pattern in the plain white shell) and she rolls those of the minor hens to the edge. The 'rejected' eggs form a frontline defence against predators such as jackals and vultures, which can eat only a few eggs or chicks. This reduces the chances of the major hen's eggs being eaten and presumably strengthens the species as the offspring of the dominant major pair are likely to be more successful than those of the minor birds. But the minor hens gain in that although they have failed to Set mates it is likely that at least some of their eggs will hatch.

By removing their eggs as they are laid, some birds may be fooled into laying way beyond the normal clutch-size. In this way a hole-nesting common flicker *(Colaptes auratus)*, a North American member of the woodpecker family, was once induced to lay 71 eggs in 73 days, also in the woodpecker family, the wryneck *(Jynx torquilla)* was made to lay 48 eggs instead of the usual 5-14. A house sparrow (*Passer domesticus*) 1 aid 51, mallards (*Anas platyrhunchos*) 80-146 and a bobwhite quail 128. Such stimulated laying is put to good use among domesticated birds such as the common chicken, which will lay for much of the year. And in protected environments, domesticated birds may be exceptionally successful with unusually large clutches. A muscovy duck once hatched 25 ducklings from a clutch of 31.

Many birds with relatively low mortality and high life expectancy lay single eggs. Some of the longest-lived birds, such as the three great albatrosses *(Diomedeidae)* and various large raptors, may breed only once in two years. The single egg laid in the summer produces a chick which is fed through the winter to fledge a year later - the lowest reproductive rate of any bird.

The redlegged partridge *(Alectoris rufa)*, on the other hand, has a most unusual method of attaining high reproductivity. It is unique among British birds at least in that one female may lay two clutches in separate nests - one for her mate to incubate and one for herself so that each will raise a brood at the same time. The average clutch size is 10-15 but up to 28 have been recorded. The high rate of egg production is offset by increased losses, for, unlike the grey partridge, the redleg never covers its eggs when left and many are taken by predators. Surprisingly, clutches left for weeks and eventually incubated by the male are hatched as successfully as those incubated immediately.

The Most Clutches, Eggs and Fledglings in One Year

The number and size of clutches, and of chicks surviving from each brood, are all governed by seasonal abundance of food. Some species articularly passerines, have a prolonged breeding season and the maximum number of broods raised in the wild is generally six. But there are reports of captive zebra finches (*Poephila guttata*) raising as many as 21 consecutive broods. In the wild this finch is an opportunist breeder, laying in dry parts of Australia at irregular intervals after rainstorms and continuing to breed until the weather deteriorates. But adults of all species are restricted by their own survival needs such as moulting and building up fat reserves for winter, and, for most, late broods do not produce as many surviving young as earlier ones.

Because of their semi-wild state and adaptation to the protecting human environment, feral pigeons (*columba livia*) often enjoy a protracted breeding season and may well hold the record for the number of clutches laid in one year by a free-flying bird, in Britain at least. However, many of their eggs do not produce fledged young. Within Britain, older, more experienced blackbirds (*Turdus merula*) probably fledge as many young as most other birds - often 17 from four broods of 4-5. Birds such as the grey and redlegged partridges (*Perdix perdix and Alectoris rufa*), which often lay over 20 eggs, generally have only one brood per season and suffer a high mortality rate. Yet the redlegged partridge may occasionally fledge more than the blackbird because it has a unique system whereby the male and female raise broods simultaneously.

Another prolific breeder is the moorhen (*Callimula chloropus*), which commonly raises 2-3 broods of 5-11 and is most expert in their care. Predation is high at the beginning of the season but there are many repeat layings as well as 'dump nesting.' Both parents share the incubation and feeding of young. The chicks leave the nest quickly - within 2-3 days - can swim and dive immediately and may even be fed by young of an earlier brood. There may be several special brood nests depending on the number of young. Some have ramps and are built soon after the chicks hatch.

The moorhen is unusual among monogamous birds in that the female plays a dominant, even aggressive, role. The largest females get the pick of the males, but, rather surprisingly, they prefer small males as these need to spend less food to keep healthy and can put on comparatively more fat. Such males need to spend less time feeding and are better placed to concentrate on incubating so that their mates can start more clutches and replace losses.

Though laid singly in the nests of different hosts, the eggs of parasitic species may produce a larger number in a year. Females of several species of African cuckoo, as well as the European cuckoo (*Cuculus canorus*), commonly lay as many as 20-24 eggs in a season.

The fastest and slowest egg laying

A few seconds to 2 hours: The actual laying, or voiding, of an egg from the end of the oviduct (vagina) is made possible by wave-like contractions of the vaginal wall and in most birds takes 1-3 minutes. However, some parasitic species such as cuckoos (Cuculidae), which must lay quickly to minimize chance of detection can lay in just a few seconds. On the other hand, turkeys (Meleagridinae) and geese (Anatidae) may take 1-2 hours.

The interval between the laying of each egg in a clutch also varies. The first egg is often laid as soon as the nest is finished, but some species start before the nest is complete and others delay for a day or two after nest completion. A regular time of day is adhered to for many species. *Songbirds* seem to prefer dawn, but pigeons and doves (Columbidae) like the early afternoon and a few species the evening. Although one egg every 24 hours is common, a three-day interval is not uncommon among birds of prey (Falconiformes) and owls (Strigiformes), for whom varying prey supply is linked to survival. When food is short the older and larger birds get the lion's share and are more likely to survive. But even slower is the masked booby (*Sula dactylatra*) with a seven-day interval between the two eggs. Where food is in short supply from impoverished waters, the booby brood is often reduced to one by sibling murder.

The Most Variably Coloured/Patterned Egg

The common guillemot's: No two birds' eggs are identical in appearance, though shape and size are remarkably similar within most species. Even in colour and pattern there is little significant variation, though in a few species great variety is a distinct advantage. Of these the common guillemot (*Uria aalge*) is by far the most notable, with a huge variety of egg shades and patterns. The ground colour ranges from white to creamy yellowish, ochreous, blue or deep blue-green, with the most extraordinary variety of markings, inter-lacing lines, sports, blotches or uniform masses of colour ranging from bright red or brown to deep black and greenish black and, occasionally, there are no markings at all. The Brunnich's guillemot (*Uru lomvia*) also has a wide variety of shell markings, but the closely related razorbill (Alea torda) has much less variable eggs.

The guillemot's great egg variation helps to prevent confusion over egg ownership in the species' large, dense colonies on the cliff ledges. It has been clearly shown that guillemots identify their own eggs on the basis of individual combinations of ground colour and marking patterns. Great-tailed grackles (*Quiscalus mexicanus*) may also recognize their eggs in this way among the crowded coconut-palm colonies of Central America.

The single guillemot egg is pear-shaped to help prevent it rolling off the cliff ledge, though many still do. Interestingly, the rolling radius changes from 17 cm (7 in) for fresh eggs to 11 cm (4 in) for fully incubated eggs, but this is a fortuitous outcome of a change in the egg's centre of gravity as the embryo develops, rather than a special adaptation. The large egg (12% of female's body weight) is rested across the feet when incubated. Its shell is thicker at the narrow end, where it is in contact with the rocky nest-ledge, and thinner at the broad end, where the chick emerges.

Parents find the right chicks in the crowded colonies through voice identification.

Eggshell is secreted in the uterus and consists largely of proteins and minerals, largely calcium carbonate in the form of calcite. Colour is derived from pigments secreted by cells in the wall of the oviduct, particularly of the uterus. The colours are deposited at different depths in the shell according to the position of the egg in the oviduct at the time of secretion. Much of the ground colour is provided by pigments in the spongy layer of the shell and must therefore be secreted at the upper end of the oviduct, while the blotches, speckles and scrawls on the surface are secreted lower down shortly before laying. The colours are derived from blood haemoglobin and bile pigments. Some shades are achieved by over-lapping of colours. Originally all eggs were probably white as the ancestral reptilian method is to bury them in sand or loose soil. Pale eggs are retained by primitive families such as cormorants (Phalacrocoracidae), pelicans (Pelecanidae) and albatrosses (Diomedeidae) and by more advanced species which nest in holes. Eggs laid away from daylight have no need for cryptic colouring. Hole-nesters, such as titmice (*Paridae*), with speckled eggs, may have taken up the habit after a period of nesting out in the open.

Parasitic species also lay a wide variety of eggs to mimic those of their hosts. The brown-headed cowbird (*Molothrus ater*) of North America and Mexico is recorded as parasitizing 206 species, some regularly. But the screaming cowbird (*Molothrus rufoaxillaris*) of South America parasitizes just one host - the bay-winged cowbird (*Molothrus badius*).

Although cuckoo species such as the European cuckoo (*Cuculus canorus*) parasitize many different hosts, individual cuckoos generally lay eggs to match just one host and therefore do not produce a wide range of egg patterns and colours. It is believed, that a female cuckoo inherits her egg colour exclusively from her mother and that the egg colouration is not affected by the father's genetic input. She chooses the right nests by seeking out the host species that reared her and is probably attracted by the habitat in which she was raised.

The Longest Incubation Periods

Wandering Albatross and Brown Kiwi - up to 85 days: Being warm-blooded, bird embryos must be kept at a constant, relatively high temperature when they leave their mother's body. Thus, except for the megapodes (*Megapodiidae*), which use incubation mounds and parasitic species such as cowbirds (*molothrus*) and cuckoos (Cuculidae), which let other birds incubate their eggs for them, all birds keep their eggs at the proper temperature by covering them so that; their body heat is transferred more or less directly to the developing embryos. The incubation period is strictly defined as the time between the laying of the last egg in a clutch to the hatching of that egg, and among species reliably studied so far the longest periods have been up to 85 days for the brown kiwi (*Apteryx australis*) and wandering albatross (*Diomedea exulans*). But that is an extreme and the incubation ranges of these two species are generally 74-84 and 75-82 days respectively. The royal albatross (*Diomedea epomophora*) has a similar range of 75-81 days, with a tendency towards the higher figure. Generally, the heavier the egg of a species, the longer the incubation: each doubling of egg weight increases incubation time by 16% on average.

Most birds develop a special brood patch for incubation. This comprises 1-3 adjacent abdominal areas which are normally relatively free of feathers all year round, but any down which occurs there, and some marginal contour feathers, are shed as the skin thickens slightly and the density of blood vessels increases.

Sometimes birds will sit on the incomplete clutch but not apply full body heat. The required temperature of 34-39°C (93-102°F), depending on species, is achieved only after a warm-up period, the length of which may be relate... tightness of sit and time required for the incubation patch to develop. Some 75% of species aid for a constant temperature of 35°C (95°F).

Most birds cover eggs for 60-80% of the incubation period, regulating the egg temperature according to atmospheric conditions. Periods spent sitting vary greatly from under an hour for many passerines such as the European robin (*Erithacus rubecula*) to several hours in most seabirds, such as gulls and terns (Laridae). Offshore-feeding seabirds such as shearwaters and storrn-petrels (Procellariidae) sit for 2-12 days at a stretch and the great albatrosses (Diomedeidae) for 2½ - 3 weeks. But the male emperor penguin's (*Aptenodyies forsteri*) 64- day continuous sitting beats them all. In some species the time spent sitting decreases as the season advances, probably due to rising air temperatures.

Maintenance of correct humidity is also important during incubation, most eggs losing about 15% of their weight before hatching. This is lost through the porous surface, which also allows important ventilation. Turning the eggs, too, is important: some species do this up to 11 times an hour to prevent 'sticking' of membranes and avoid the adverse effects of any temperature gradients in the nest.

In 54% of species sexes share the incubation. For 25% it is the female alone, for just 6% the male alone, and it varies between male, female and both in about 15%. Single birds sit more continuously, but where both sit hatching is not necessarily quicker. Some parents, such as bushtits (*Psaltriparus*) of North and Central America, are assisted in the incubation by adults without nests of their own. Others, such as anis (*Crotophaga*) and acorn wood-peckers (*Melanerpes formicivorus*), have communal nests with several females laying and males and females sharing the incubation. The waxbills (*Estrilda*) are the only species in which the male and female incubate side by side on the nest.

The percentage of egg volume taken up by the yolk food supply at the start of incubation depends upon how advanced the chicks need to be at hatching. For most passerines, which hatch blind, naked and helpless, a 12-15-day incubation is normal, with the yolk taking up only about 20% of the egg.

So-called precocial species (hatching young which are soon able to leave the nest) such as shore-birds, ducks and gamebirds, which hatch in 3-7 weeks, require a larger yolk of about 35% total egg volume. Such chicks emerge from the shell with a full coat of protective down and are soon able to fend for themselves.

An embryo consumes the yolk until just before hatching, when the remains of the yolk sac pass into the body of the chick through the umbilical opening and, in some instances, may continue to nourish the bird for several days after hatching, giving the parents or chick time to establish the new feeding routine.

The Shortest Incubation Periods

Small passerines- 10 days: Ten days seems to be the shortest possible incubation period for a fairly small number of passerines which lay eggs weighing under 1 g (0.035 oz). These include the hawfinch (*Cocothraustes cocothraustes*), blackcap (*Sylvia atricapilla*), lesser whitethroat (*Sylvia curraca*), redpoll (*Acanthis flarnmea*), great-spotted woodpecker (*Picoides major*) and the black-billed cuckoo (*Coccyzus erythropthalmus*). But these birds may also incubate for 11 days or even longer, according to how long the egg was in the oviduct and whether or not it was fully formed when laid.

Many more small birds have 12-14 day incubation periods, their young generally hatching form eggs with relatively small yolks. Birds which hatch in a moon advanced state are sustained by proportionately larger yolks over longer incubation periods.

The Fastest and Slowest Hatching

30 minutes to 6 days: The actual process of 'escaping' from the egg

— hatching — can take anything from 30 minutes or so in small passerines to six days for the larger albatrosses (Diomedeidae). But for most small to medium-sized birds it takes from a few hours to most of a day on average. In most clutches the eggs do not hatch together, but in sequence, reflecting the intervals in laying. Preparation for hatching begins quite early in the incubation. As the embryo's moisture gradually evaporates through the porous eggshell an air pocket gathers at the blunt end of the egg, between the inner and outer shell membranes. The developing chick uses this for breathing over several days prior to hatching, though it still depends on a part of the embryo called the allantois for oxygen exchange.

The actual 'tools' or hatching — the temporary egg tooth on the tip of the upper mandible, and the special hatching muscle — also form early in the incubation, and the eggshell becomes conveniently weaker as it provides minerals for the developing chick's skeleton.

First the embryo reafigns its body to lie along the length of the egg, the best position to force its way out from the blunt end. Convulsive thrusting forces the bill against the shell, and with the aid of the horny egg tooth the shell is soon cracked or dpiped'. After the chick has broken out the hatching muscle withers away and the egg tooth either drops off or is reabsorbed into the bill, according to species. From pipping to full emergence is the hatching period. During this, the chicks of some species are helped by the parents, which may pick at the shell or poke into the cracks. It is said that ostriches crack their eggs with their breastboiies,aiid even drag the chicks out with their bills. Some parents will begin bringing food in response to the chick's peeping even from inside the unbroken shell.

Quickest to Fly

Cabot's Tragopan and Megapodes - within 24 hours of hatching: It would not be right to say these species are the fastest at learning to fly because they seem to be born able to fly without any training or exercising of flight muscles. Most birds are born with down and are said not to be fledged until they have acquired their flist true contour feathers, including the primaries which will propel them in flight. This generally takes at least 10 days and ground-nesting species, whose young leave the nest soon after hatching, are usually much quicker at getting into the air than birds which are born naked and helpless and remain in the nest after hatching. This is taken to the extreme, in the megapodes (Megapodiidae), whose young hatch in special incubation mounds and must fend for themselves immediately. They never see their parents and instead of the usual down they are born with extremely advanced contour-like feathers which enable them to fly and escape from predators almost at once.

Equally advanced on hatching is Cabors tragopan (*Tragopan caboti*) of China a member of the pheasant family. Its chicks hatch with a thick coat of coarse, shaggy down, but with primaries so far advanced they are immediately able to flutter up to perches. They also climb well and within a day or two can fly as well as any young passerine. Yet, although incubation is generally long among pheasants, it is not particularly long for this advanced bird, being just 28 days. This species is also unusually arboreal and lives in dense forests with very thick undergrowth which is often saturated. In this habitat a well- developed plumage is a distinct advantage.

Cabot's tragopan is listed as endangered, but reserves have recently been established within its historic range. Only one nest has ever been described from nature and that was 9m (30 ft) up in an old squirrel drey.

The satyr tragopan (*Tragopan satyr*) is also highly precocial at hatching and within two or three days can fly up to an elevated perch to roost under its mother's wings. It too nests in trees, and is widely hunted and trapped with nooses in Nepal, legal protection being little help.

Temminck's tragopan (*Tragopan temmincki* — the Chinese crimson horned pheasant) of the eastern Himalayas can also fly up to elevated perches within two or three days of hatching. Yet the chicks grow slowly, are sensitive to disease and chilling and do not attain full adult male plumage until the second year, the adult female plumage coming first. This tragopan has a wider range and is more secure, but it too is extensively trapped and shot for its feathers and flesh, and much of its forest habitat is being cut for timber or razed for agriculture.

The shortest fledging time for altricial young — those helpless when hatched — is about eight days for some of the warblers (Sylvinae) and finclies (Fringillidae).

Slowest to Fly Great

Albatrosses - 9-12 months: The slowest birds to fledge are the great albatrosses (Diomedeldae), especially the royal (Diomedea epomophora) and the wandering (*Diomedea exulans*), which generally take 9-12 months.

At first the bulky young birds sit very quietly in the nest. They have two successive coats of down and are brooded by a parent for the first three to five weeks , after which they are visited and fed at intervals on regurgitated food. As the time approaches to leave they fidget and flap. Most fly straight from the nest, especially if inland, but they are likely to make abortive flights and land clumsily in unsuitable places. Sometimes they are drawn to village lights, and late-developing young may swim out to sea if they can reach the water. But strong chicks fly well-immediately and never see their parents or birthplace again. When fledging is as long as this the parents can breed only once in two years.

Britain's slowest bi:d to fledge is the mute swan (*Cygnusolor*), which leaves the nest after a day or two but does not fly for some four months. The cygnets are tended by both parents.

The Fastest to Breeding Maturity

Quails - front 5 weeks: The five species of quail in the genus *Coturrtix*, including the common quail (*Colurnix cottirnix*), reach maturity before any other bird — one reason why they have been bred in large numbers for the table. They are able to reproduce from as little as five weeks, through in most quails adult plumage is not attained before 10-12 weeks. Some other bird species are also capable of reproduction while still in immature sub-adult before they are sexually mature.

Most small songbirds breed for the first time-when just under one year old, in their second summer, and many others near the end of their second year. But some individuals within a species mature as year earlier than others.

Although physiologically capable of breeding, some species delay breeding for years as they need to develop their food- finding skills before they are capable of feeding themselves and a family. Hunting species in particular rely very much on experience, so young predatory birds continue to depend on their parents while practising their skills. Some manage this in two o: three weeks but the African crowned eagle (*Stephanoaetus coronatus*) continues to receive food from its parents for up to a year even though it can fly.

The Slowest to Breeding Maturity

Albatrosses — from 6-10 years old: Despite the fact that the larger albatrosses (Diomedeidae) — notably the royal (*Dioniedea epomophora*) and the wandering (*Diomedea exulans*) - have very long incubation and fledging periods, they still delay for the record period of 6-10 years before breeding for the first time. There are even reports of a first breeding at 12 years old, through the birds are physiologically capable of reproducing before this. And even after all this time first efforts are frequently unsuccessful.

After fledging, the transequatorial migrants set out on long journeys which must include a period of starvation as they cross the tropics. The early years are spent at sea, and they sometimes gather in nurseries where there is a good food supply. As they get older they return to land for increasing periods wandering in search of new nest-sites, though most return to their birthplaces. Further years are spent displaying and excavating nests before they even attempt to breed.

Index